SPECIAL THEORY OF RELATIVITY

S. P. Puri

Former Professor
Panjab University
Chandigarh

Pearson

Assistant Editor–Acquisitions: Jigyasa Bhatia
Editor–Production: Nikhil Rakshit

ISBN: 978-81-317-8501-0

First Impression

Published by Dorling Kindersley (India) Pvt. Ltd., lincencees of Pearson Education in South Asia.

Head Office: Pearson Education, 7th Floor, Knowledge Boulevard, A-8 (A), Sector-62, Noida, India
Registered Office: 11 Community Centre, Panchsheel Park, New Delhi, 110017, India

Compositor: County Caramels
Printed in India at Repro India Limited

Contents

Preface vii
Historical Perspective ix

1. Newtonian Mechanics and Galilean Principle of Relativity 1

1.1 Difficulties of Pre-relativity Physics. Galilean Transformation 1
1.2 Newton's Laws and Inertial Frames of Reference 4
1.3 Absolute Time and Absolute Space 5
1.4 Wave Equation Under Galilean Transformation 6
1.5 Search for a Universal Frame of Reference 8
References 15
Problems 16

2. Lorentz Transformation and Its Kinematic Consequences. Intervals and Causality 17

2.1 Einstein's Postulates of Special Theory of Relativity. Principle of Relativity and Constancy of Velocity of Light 17
2.2 Lorentz Transformations. Axiomatic Derivation 20
2.3 Kinematic Consequences of Lorentz Transformation 28
2.4 Geometric Interpretation of Lorentz Transformation 37
2.5 The Light Cone 40
2.6 Transformation of Velocity and Acceleration 42
References 45
Problems 46

3. Mathematical Background 49

3.1 Linear Orthogonal Transformation and Its Invariants 49
3.2 Minkowski's Four-dimensional Space-Time World 53
3.3 Lorentz Transformation as Rotation in Four-space 55
3.4 Four-vectors 60
3.5 Tensor Algebra. Anti-symmetric Four-tensor 65
3.6 Group Property of Lorentz Transformation 68
References 73
Problems 73

4. Relativistic Mechanics of a Particle. Collisions and Conservation Laws 75

4.1 Relativistic Lagrangian for a Free Particle 75
4.2 Energy and Momentum of a Free Particle 76
4.3 The Conservation of Momentum. Variation of Mass with Velocity 81
4.4 Four-dimensional Velocity and Acceleration 88
4.5 Relativistic Equation of Motion. Minkowski Force 90
4.6 Energy–Mass Equivalence 94
4.7 Centre of Momentum Transformation 106
4.8 Kinematic Relations for Collisions 110
References 114
Problems 114

5. Optical Applications of Lorentz Transformation 117

5.1 Elements of Wave Propagation 117
5.2 Frequency and Wave-vector Transformations 120
5.3 The Relativistic Doppler Effect. Aberration 121
5.4 Reflection of Light by a Moving Mirror 126
5.5 Propagation of Light in a Moving Medium: Fresnel's Dragging Coefficient 126
References 128
Problems 128

6. Covariant Electrodynamics 131

6.1 Invariance of Maxwell's Equations Under Lorentz Transformation 131
6.2 Transformation of Electromagnetic Field Intensities 134
6.3 Electromagnetic Field Tensor 137
6.4 Covariant form of Maxwell's Equations 144
6.5 Lagrangian and Hamiltonian for a Charged Particle in an Electromagnetic Field 147
6.6 Equation of Motion of a Charge in an Electromagnetic Field 151
6.7 Maxwell's Equations by Variational Technique 157
6.8 Electromagnetic Energy–Momentum Tensor 160
6.9 Field of a Charge in Uniform Motion 165
6.10 Retarded Potentials. Lienard–Wiechert Potentials 168
6.11 Field Strengths of an Accelerated Charge 171
Problems 176

7. Applications of Special Theory of Relativity 179

7.1 Experimental Verification of the Special Theory 179
7.2 Thomas Precession: Spin–Orbit Interaction 183

7.3 Relativistic Classification of Particles 188
7.4 Zero Rest-mass Particles: Photons and Neutrinos 190
7.5 De Broglie Waves for a Free Particle 194
7.6 The Uncertainty Principle 197
References 202
Problems 204

8. Introduction to General Relativity 205

8.1 Equivalence of Inertial and Gravitational Mass 206
8.2 The Principle of Equivalence 208
8.3 Einstein Elevator 208
8.4 The Principle of Covariance 210
8.5 Gravitational Effect on Clocks 211
8.6 Classical Tests of General Relativity 212
8.7 The Renaissance of General Relativity 214
References 214
Problems 215

Bibliography **217**

Index **218**

Preface

This book grew out of lectures delivered to postgraduate students in physics at the universities of Roorkee, Roorkee and Panjab, Chandigarh since 1961. *Special Theory of Relativity* is primarily intended as a textbook for the physics and mathematics students at the B.Sc. level in the Indian universities. There has been resurgence in subjecting the postulates as well as predictions of special relativity to greater precision tests since 1959, and the book is an up-to-date exposition of the subject-matter. The purpose of the book is twofold: to develop the basic ideas of the special theory of relativity and to indicate the influence of such a modification of the usual concepts of absolute space and time of Newtonian mechanics, on the structure of physical theory. Special theory adapted the concept of inertial frame to the basic postulate of the constancy of the velocity of light in vacuum. Furthermore, it implies that the velocity of light is not just the speed of propagation of electromagnetic radiation but a fundamental feature of the manner the space and time are unified as space-time.

After a brief introduction related to the difficulties of pre-relativity physics, the book deals with the failure of Galilean transformations to preserve the invariance of Maxwell's equations, thus asserting the need for more general transformations. Further, Lorentz transformations are deduced axiomatically from the principle of the invariance of the velocity of light as well as an orthogonal rotation in Minkowski space. The kinematical and dynamical consequences of Lorentz transformation are presented with the help of three-dimensional calculus as well as the four-dimensional formulation. The available experimental evidence is compared with the predictions of the theory.

The necessary algebra of orthogonal transformations, four-vectors and tensors, is presented and illustrated by means of numerous solved examples, before the quasi-geometrical language provided by the four-dimensional space-time continuum is used in subsequent development of the subject-matter of dynamics and electromagnetism. Relativistic optics is also presented in detail.

Vacuum electrodynamics is formulated in a covariant manner and Maxwell's equations are derived by variational technique. In addition, the energy-momentum tensor for electromagnetic field is derived. Lastly, it is shown that the wave-particle duality emerges from relativistic considerations. A brief introduction to general relativity constitutes the finalé to the book.

As distinguishing features of the book, I would like to mention the following:

1. It is an up-to-date exposition of the subject-matter of special theory of relativity, with detailed references to the precision tests of its postulates, predictions and Lorentz invariance.
2. It accounts for the formulational as well as the applicational aspects of the subject.
3. The style of presentation is lucid, and direct and sufficient mathematical details are included so as to sustain the interest of the reader.
4. With a view to augmenting the usefulness of the book for self-study, a large number of solved problems have been included in addition to the chapter-end problems.
5. Apart from Lorentz–Fitzgerald length contraction where both the observers have to collaborate, the observations of a single observer have been discussed.
6. Lorentz transformations have been derived from the group property of the transformation.

I express my sincere appreciation to Prof Vishwamittar for his critical reading of the manuscript and making useful suggestions which led to the improvement and clarity of presentation. I am beholden to him.

Lastly, my special thanks is due to my wife, Ramesh, whose deep interest and constant encouragement served as a beacon of hope throughout.

Any suggestions or comments to improve the subject matter will always be welcome.

—S. P. Puri

Historical Perspective

The romance of creative achievement lies in reciting the story itself, as how it finally evolved historically. The earliest notions of space were that of Aristotle who visualized the world as an absolute space. You could easily infer whether you are at rest or moving. Contrary to Aristotle, Heractilus held that all things are in motion or in a constant state of flux though sometimes not visibly so. This concept of dynamic equilibrium is surprisingly a modern view. But Aristotle's views held sway for a number of centuries. Around the time of renaissance, Tycho Braha made accurate measurements of the motion of planets in our solar system. Kepler based his analysis on this data and formulated three laws of planetary motion, the first of which states that planets moved in elliptical orbits. Galileo Galilei invented the telescope and made some significant advances in astronomy; he came up with five laws of motion and first known modern statement of the *principle of relativity*, based on an infinite velocity of propagation of interactions. According to Galileo's principle of relativity, the equations expressing the laws of nature are invariant with respect to transformations of coordinates and time from one inertial frame to another.

Isaac Newton developed three laws of motion, in his work *Principia*, in 1687, that relied on Galileo's principle of relativity. Time and space were considered absolute for all observers moving in uniform relative motion. This seventeenth-century theory held its sway until the nineteenth century when electric and magnetic phenomena were studied theoretically.

Maxwell showed that the velocity of electromagnetic waves or any disturbance in the fields will propagate with velocity c, which is the ratio of the electromagnetic to electrostatic unit of electrical charge. The value of c as found experimentally is 2.99792×10^8 m/s, which agrees with the velocity of light in *vacuo*. This agreement of velocity caused Maxwell to postulate that light itself was an electromagnetic phenomenon. It was quite natural to postulate a medium for the transmission of light like sound. Such a medium was called ether. This medium had to be incredibly stiff for the light to move so fast in it, and at the same time it should be soft and insubstantial; such a medium has never been detected.

Maxwell's equations do not depend on the speed of the source of light. This was unlike, say, throwing a ball out of a moving car. It was thought that now we have a way of measuring the speed of earth through the ether. We just have to set up a light source and a meter stick and see the difference when we point it in the direction of earth's movement compared to some other direction. The speed of light is independent of the speed of light source but will depend how fast the observer is moving through the medium.

Prompted by Maxwell's ideas, Michelson carried out his famous experiment in 1881 and it showed no significant difference in the speed of light. Michelson reported:

The result of the hypothesis of a stationary ether is shown to be incorrect and the necessary conclusion follows that the hypothesis is erroneous.

Michelson–Morley repeated the experiment in 1887 and again reported that no effect had been found. It appeared that the velocity of light was independent of the velocity of the observer. They refined their experiment and repeated it many times up to 1929, but with the same outcome.

It was in 1887 that Vöigt, while working on Doppler shift, wrote down the following equations:

$$x' = x - vt, \quad y' = \frac{y}{g}, \quad z' = \frac{z}{g}, \quad t' = t - \frac{vx}{c^2}$$

and showed that certain equations preserved their form under these transformations. With a different scale factor, these transformations are known as the Lorentz transformations. The group of Lorentz transformations gives the geometry of special relativity.

Irish physicist George FitzGerald in 1889 published a paper, *The ether and the earth's atmosphere*, and proposed an ad hoc explanation of the negative results of Michelson–Morley experiment. It stated:

The length of material bodies changes according as they are moving through the ether or across it, by an amount depending on the square of the ratio of their velocities to that of light.

Unaware of FitzGerald's paper, Lorentz in 1892 proposed an identical contraction of the moving object along the line of motion and was quick to acknowledge that it was proposed first by FitzGerald while explaining the negative results of Michelson–Morley experiment.

In 1898, Larmor in his article "*ether and Matter*," wrote on Lorentz transformations and showed that FitzGerald contraction followed as a consequence. Thus, Larmor was the third person to write down the Lorentz transformations, which are an improvement over Galilean transformations.

The concept of ether as a material substance was questioned by Paul Drude, who in 1881 wrote:

The concept of an ether absolutely at rest is the most simple and the most natural—at least if the ether is conceived not as a substance but merely space endowed with certain physical properties.

Henri Poincaré, in his address to the Paris Congress in 1900, raised the question: 'Does the ether really exist?' In 1904, in his address to the International Congress of Arts and Science in St Louis, Poincaré suggested that old assumptions were wrong and there should be no way at all to tell whether you were moving or at rest or how fast you are moving except relative to something else and so resurrected the principle of relativity, and stated:

The impossibility of demonstrating absolute motion is a general law of nature.

Einstein's first paper on special relativity was received on 30 June 1905 and is marked for its beauty and simplicity. He did not attempt to explain the experimental results, but straightforwardly stated:

The introduction of a light-ether will prove to be superfluous, according to the view to be developed in the article. Neither will a space in absolute rest endowed with special properties be introduced nor will a velocity vector be associated with a point of empty space.

He introduced inertial frames which are in uniform motion with respect to each other. Special relativity is based on the following two postulates:

1. The laws of physics assume the same form in all inertial frames.
2. In any inertial frame, the velocity of light c is the same and is independent of the motion of the source.

From these two postulates, Einstein deduced Lorentz transformations and like Poincaré proved the group property. The FitzGerald contraction follows in a straight manner. He also mentioned the clock paradox in his paper.

In September 1905, Einstein deduced the mass-energy equivalence relation:

$$E = mc^2$$

Max Planck wrote a paper on relativity in 1908 and his eminence gave a fillip to the acceptance of it. Also in 1908, Minkowski derived Maxwell–Lorentz equations for electromagnetic field in tensor form.

As is clear, the ideas of relativity were developed by many people. The basic principle was from Galileo, which was embraced by Newton and then restated and refined by Poincaré. The Lorentz transformations were derived by Vöigt, Larmor, and Lorentz. The experimental evidence was first provided by Michelson–Morley and later by others. The main contributors to special relativity were undoubtedly Lorentz, Poincaré, and of course, the founder of the theory, Einstein.

In 1912, Lorentz and Einstein were jointly proposed for a Nobel Prize for their work on special relativity. They were recommended by Wien, the 1911 winner. The Nobel committee was at first cautious and waited for the experimental confirmation of the theory, and Einstein never received the prize for relativity. Meanwhile, Einstein moved further to work on momentous contribution of *general relativity.*

1

Newtonian Mechanics and Galilean Principle of Relativity

Albert Einstein formulated the special theory of relativity in 1905, in the paper[1] '*On the Electrodynamics of Moving Bodies*,' after considerable contributions by Hendrik Lorentz and Henri Poincaré. It is the physical theory of measurement in inertial frames of reference and generalizes Galileo's principle of relativity—that all uniform motion is relative and there is no absolute state of rest. Furthermore, the special theory incorporates the principle that the speed of light is the same for all inertial observers, irrespective of the state of motion of the source. The theory is termed special since it applies the principle of relativity only to inertial frames and not to accelerating frames or in gravitational fields. Later, Einstein developed the general theory which is applicable to any frame and it includes the effects of gravity.

According to special theory, Newtonian notions of absolute time and space are untenable and are perceived differently by observers in different frames of reference. Such counterintuitive concepts have been verified experimentally. Furthermore, space and time have been made non-separable as space–time continuum. However, the predictions of special relativity agree with Newtonian mechanics in the realm of velocities small compared with the speed of light.

Special relativity indicates that c is not just the speed of propagation of electromagnetic radiation but a fundamental feature of the manner the space and time are unified as space–time. The consequence of it is that a particle having mass cannot be accelerated to the speed of light.

1.1 DIFFICULTIES OF PRE-RELATIVITY PHYSICS. GALILEAN TRANSFORMATION

Newtonian or classical mechanics is based on the assumption of instantaneous propagation of interactions and embodies the concepts of absolute space and absolute time, and its predictions are borne out by experimental facts in the realm of velocities far small compared with the velocity of light. However, when the velocities involved approach the velocity of light, some basic alterations in these concepts become inevitable, so that these could explain the experimental facts. Before we undertake to find the changes in the formulation of classical mechanics, let us examine in greater detail the limitations that confronted the pre-relativity physics.

According to Newton's first law of motion, a body continues to be in a state of rest or of uniform motion in a straight line unless it is impelled by an outside force to change that state. This law is also called the *law of Inertia* and the coordinate system in which this law is valid is called the *inertial system*. If two systems of reference are moving uniformly relative to each other and if one of these is an inertial system, then clearly the other is also inertial. In order to see the inadequacy of a mechanical experiment to distinguish between two observers A and B as to which is at rest and which is in motion, when they are moving with velocity v relative to each other, let us consider the following:

A and B are two observers moving with velocity **v** relative to each other. Take rectangular axes OX, OY, and OZ fixed with reference to A (frame S) and parallel axes $O'X'$, $O'Y'$, $O'Z'$ fixed with reference to B (frame S'), Figure 1.1. Let the origin of the primed axes be (vt, 0, 0), which in the unprimed axes implies that the motion is in the OX or $O'X'$ direction and at $t = 0$, the origins O and O' coincide.

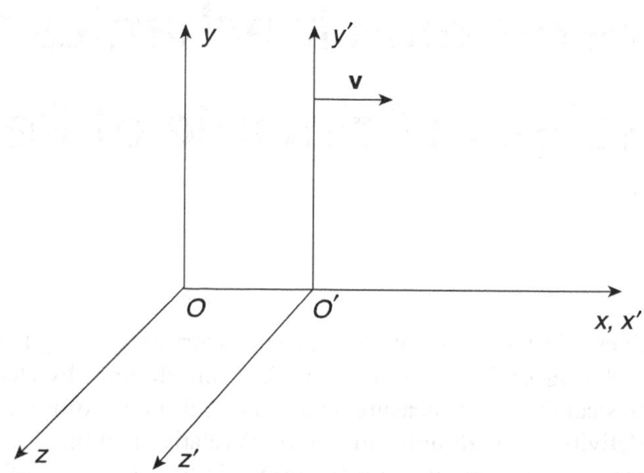

Figure 1.1 *The primed system is moving with velocity* **v** *relative to the unprimed system.*

Suppose P is a point in space, described by an observer at O in S by coordinates x, y, z at time t and that by observer at O' in S' by coordinates x', y', z' at time t'. All physicists before 1900 asserted that

$$x' = x - vt \qquad [1.1(a)]$$

$$y' = y \qquad [1.1(b)]$$

$$z' = z \qquad [1.1(c)]$$

$$t' = t \qquad [1.1(d)]$$

These equations are known as the *Galilean transformation equations*.[1]

The reference frames moving uniformly and rectilinearly and relative to one another are equivalent. The frames S and S' differ only in the velocity of S' relative to S being equal to **v***, whereas the velocity of frame S relative to S' is equal to –**v**. Therefore, one can obtain the reverse transformations by changing primed into unprimed quantities and changing the sign of v from v to $-v$ and vice versa.

Thus,

$$x = x' + vt' \qquad [1.2(a)]$$

$$y = y' \qquad [1.2(b)]$$

$$z = z' \qquad [1.2(c)]$$

$$t = t' \qquad [1.2(d)]$$

These equations can be derived algebraically from Eqs [1.1(a)]–[1.1(d)].

---

* Bold face type symbols are used to represent vector quantities. The same symbol in italics will represent only the magnitude of the vector quantity.

Let us examine one of the consequences of the Galilean transformation. Imagine that two events occurred in the frame S on the x-axis at the point x_1 at the time t_1 and the other at x_2 at time $t_2 (t_1 \neq t_2)$. Is it possible to select the frame S' in which both the events occur at the same point in space? The x-coordinates of the events in S' are

$$x_1' = x_1 - vt_1 \quad [1.3(a)]$$

$$x_2' = x_2 - vt_2 \quad [1.3(b)]$$

Thus,

$$x_2' - x_1' = x_2 - x_1 - v(t_2 - t_1) \quad (1.4)$$

To make $x_2' - x_1' = 0$, we have

$$v = \frac{(x_2 - x_1)}{(t_2 - t_1)} \quad (1.5)$$

Thus, during the period $(t_2 - t_1)$, the frame S' brings the point x_1' to the point where the second event took place. However, this was possible because according to classical mechanics v can have any magnitude. In the theory of relativity, the velocity of a reference frame (like the velocity of a material particle) is limited. The required frame S' is far from being always realized.

The equations of motion of a particle of mass m subject to the force $\mathbf{F}(F_x, F_y, F_z)$ are given by Newton's law as

$$m\frac{d^2x}{dt^2} = F_x \quad [1.6(a)]$$

$$m\frac{d^2y}{dt^2} = F_y \quad [1.6(b)]$$

$$m\frac{d^2z}{dt^2} = F_z \quad [1.6(c)]$$

Let us transform the left-hand sides of Eqs [1.6(a)]–[(1.6(c)] under the transformations of Eqs [1.2(a)]–[(1.2(d)]. These become,

$$m\frac{d^2x'}{dt'^2} = F_x \quad [1.7(a)]$$

$$m\frac{d^2y'}{dt'^2} = F_y \quad [1.7(b)]$$

$$m\frac{d^2z'}{dt'^2} = F_z \quad [1.7(c)]$$

Thus, the Galilean transformations predict that the equations of motion are invariant (preserve their form) in going from one system to another moving with a uniform velocity with respect to one another, none of which is unique or pre-eminent. Neumann[2] called this set of inertial frames by the name *The Body Alpha*. Thus, "The Body Alpha" had replaced the concept of absolute space, but that of absolute time had been left intact.

Further, the transformations (Eqs [1.1(a)]–[1.1(d)]) predict that the velocity of light should be different in the two systems. In other words, the Newtonian law of addition of velocities predicts that the velocity of light is different in different directions and depends upon the relative uniform motion of the observer, the medium, and the source. If c is the velocity in the unprimed system of a point on the wave surface emitted by a source at the origin, then

$$\mathbf{r} = c\mathbf{n} \tag{1.8}$$

where $\mathbf{n}$ is a unit vector along r. Transforming the above equation to the primed system, we get

$$\mathbf{r}' = c\mathbf{n} - \mathbf{v} \tag{1.9}$$

Thus the wave velocity in the primed system is no longer c and since it depends on the direction, the waves are not spherical.

Classical mechanics tacitly assumes that the interactions propagate instantaneously at infinite velocity. The forces exerted on a particle by the others at a particular instant of time depend solely on the positions of the particles at this instant. But the velocities with which we have to deal usually are so small that the velocity of interactions or *signal velocity* when taken to be infinite, does not affect the accuracy of the results. These ideas were embodied in the Galilean principle of relativity, based on an infinite velocity of propagation of interactions, according to which the equations expressing the laws of nature are invariant with respect to transformations of coordinates and time from one inertial system to another. Although the Galilean principle of relativity pertains exclusively to mechanical phenomena, it was the first step towards the principle of relativity which is applicable not only to mechanics but to all physics.

1.2 NEWTON'S LAWS AND INERTIAL FRAMES OF REFERENCE

The Newton's three laws of motion are the bedrock of Newtonian mechanics and are stated as follows. Of these, the first (Galileo's law) is really a special case of the second. These are as follows:

1. A body at rest or in motion, subjected to no impressed force, moves due to inertia, i.e., uniformly and rectilinearly.
2. The force on a particle equals the product of its mass and acceleration: $\mathbf{f} = m\mathbf{a}$.
3. The forces of action and reaction are equal and opposite, e.g., if a particle A exerts a force $\mathbf{f}$ on a particle B, then B reacts with a force $-\mathbf{f}$ on A. (The assumption of absolute time, i.e., of absolute simultaneity, is tacit if action at a distance principle is implicit.)

Physical laws are stated relative to some reference frame, with respect to which one can specify physical quantities like velocity, acceleration, force, etc. Among the frames of reference, rigid ones are the preferred ones and among these are the inertial frames. The Newtonian laws of dynamics hold in the inertial frames.

One can specify the position of one body only relative to another. A reference frame is the space determined by a rigid body regarded as a base. One can imagine the extension of the rigid body as far as desired by a lattice of measuring rods. A point is located in space by knowing its three coordinates with respect to the origin of the reference system.

The state of rest or of motion of a body is specified with reference to a frame, the former refers to the situation when the body occupies the same position in the reference system, whereas the latter implies a change of position. Relativistically, one includes timekeeping in addition to position measuring in the reference frame. For this purpose, one presupposes identical clocks at the lattice points distributed throughout the surrounding space.

Newton's first law enables to single out inertial frames among rigid frames: a rigid frame is called inertial if free particles move without acceleration relative to it. Since gravity exists between any two particles, no particle near a heavy mass is (force) free. Thus, the concept of inertial frame will be valid in regions of the frame which are sufficiently removed from the attracting matter.

The frame of fixed stars was recognized as inertial to considerable accuracy and was regarded as inertial frame for Newton's laws for a long time. Now when our universe is known to expand and our galaxy is known to rotate, instead of 'fixed stars' we consider that rigid frame as inertial in which all other galaxies appear to recede radially. Such a rigid frame is symmetrical relative to the distant universe.

Two frames of reference can be said to be inertial frames of reference with respect to one another when they are either at rest or in uniform relative motion. Just as there is no absolute motion, there is no absolute space or absolute reference frame.

Another property that can be utilized for defining inertial frames is the one according to which the equation of motion of a body takes on the simplest form, in the sense that it is free from certain additive terms that are characteristic of frames accelerated with respect to the inertial frames. The additional terms, called *inertial forces*, arise from the rotation of the frame or translational acceleration. The accelerated frame and the uniformly rotating frame of reference are cases of non-inertial frames.

A local inertial frame is defined as a reference frame in which a body shielded from all external influences has zero acceleration. These criteria are achieved by assuming that the local inertial frame is in free fall in the existing gravitational fields, the gravitational field due to the material content of the frame itself being zero. In addition, the frame has zero spin relative to the fixed stars. An orbiting satellite without spin and free of drag could be idealized as a local inertial frame. If a bullet is fired inside such a satellite, it will move with constant velocity in a straight line relative to the cabin.

1.3 ABSOLUTE TIME AND ABSOLUTE SPACE

It is worthwhile to dwell on the Newtonian ideas of space and time since these correspond to our everyday experience, and the transition to the concepts of space and time according to the special theory of relativity leads to the renunciation of the classical views. Furthermore, a still further step from these classical ideas was taken by Einstein in his general theory of relativity.

According to Newton's *Philosophiae Naturalis Principia Mathematica* (1687), these classical ideas of Newton are as follows:

Absolute space, in its own nature, without relation to anything external, remains always similar and immovable.

It is impossible to select 'motionless absolute' space, just on the basis of mechanical experiments, since equation of motion is invariant under Galilean transformation. Such a search for absolute frame was considered possible on the basis of optical phenomena. It was presumed that Newton's privileged selected reference frame, that is motionless absolute space, is the heliocentric frame. But it became clear that there is no privileged frame at all, rather there is a whole class of privileged frames, called inertial frames, in which the laws of physics assume simple form.

About time, Newton wrote:

Absolute, true and mathematical time, of itself and from its own nature, flows equally without relation to anything external, and is otherwise called duration.

Thus, according to Newton, space and time exist by themselves and do not depend on material objects located in space.

Newton's concept of absolute space has never lacked critics, from his contemporaries, Huygens, Leibniz, and Bishop Berkeley to Earnest Mach in the nineteenth century to Einstein in the twentieth century. The cogent arguments against the existence of absolute space include the following:

1. It is purely *ad hoc* and explains nothing.
2. There is no unique way of locating Newton's absolute space within an infinite class of inertial frames.
3. 'It conflicts with one's scientific understanding to conceive of a thing, which acts but cannot be acted upon.' The words are of Einstein, but he attributes the thought to Mach. It questions not only absolute space but also the set of all inertial frames.

The classical mechanics based on these concepts of absolute space and absolute time and the assumption of instantaneous propagation of interactions and its predictions are borne out by experimental observations in our everyday life. The reason for its success rests with the fact that these concepts are approximately correct and departures from them are insignificant when the velocities involved are far less than the velocity of light in vacuo. But when the velocities involved approach the velocity of light, such as for elementary particles, some basic alterations become essential in these concepts for explaining the experimental facts.

To sum up, it may be stated that according to present day viewpoint, there is no absolute space and consequently no absolute motion. All inertial frames are equivalent. The special theory of relativity shows that time readings of an event is different in different frames of reference. Therefore, time reading is dependent on the state of motion. The general theory of relativity by Einstein implies that properties of time and space are not prescribed for ever but depend on the objects located in space.

1.4 WAVE EQUATION UNDER GALILEAN TRANSFORMATION

Maxwell's electromagnetic equations summarize in differential form the following basic laws of electromagnetism (in Gaussian units).

Faraday's law:
$$\nabla \times \mathbf{E} = -\frac{1}{c}\frac{\partial \mathbf{B}}{\partial t} \qquad [1.10(a)]$$

Ampere's law for time-dependent fields:

$$\nabla \times \mathbf{H} = \frac{1}{c}\frac{\partial \mathbf{D}}{\partial t} + 4\pi \mathbf{J} \qquad [1.10(b)]$$

Absence of free magnetic poles:

$$\nabla \cdot \mathbf{B} = 0 \qquad [1.10(c)]$$

Coulomb's law:

$$\nabla \cdot \mathbf{D} = 4\pi\rho \qquad [1.10(d)]$$

where **E** and **H** are electric and magnetic field intensities respectively; **D** and **B**, the electric displacement and magnetic induction vectors; ρ, the charge density; and **J**, the vector current density. In free space, we have

$$\mathbf{D} = \mathbf{E} \tag{1.11(a)}$$

$$\mathbf{H} = \mathbf{B} \tag{1.11(b)}$$

and in the absence of charges and currents, Eqs [1.10(a)]–[1.10(d)] become

$$\nabla \times \mathbf{E} = -\frac{1}{c}\frac{\partial \mathbf{B}}{\partial t} \tag{1.12(a)}$$

$$\nabla \times \mathbf{B} = -\frac{1}{c}\frac{\partial \mathbf{E}}{\partial t} \tag{1.12(b)}$$

$$\nabla \cdot \mathbf{B} = 0 \tag{1.12(c)}$$

$$\nabla \cdot \mathbf{E} = 0 \tag{1.12(d)}$$

Now,

$$\begin{aligned} \nabla \times \nabla \times \mathbf{B} &= \nabla(\nabla \cdot \mathbf{B}) - \nabla^2 \mathbf{B} \\ &= \frac{1}{c}\frac{\partial}{\partial t}(\nabla \times \mathbf{E}) \\ &= -\frac{1}{c^2}\frac{\partial^2 \mathbf{B}}{\partial t^2} \end{aligned} \tag{1.13}$$

so that the vector **B** satisfies the wave equation

$$\nabla^2 \mathbf{B} - \frac{1}{c^2}\frac{\partial^2 \mathbf{B}}{\partial t^2} = 0 \tag{1.14}$$

Similarly, we obtain the result for **E**:

$$\nabla^2 \mathbf{E} - \frac{1}{c^2}\frac{\partial^2 \mathbf{E}}{\partial t^2} = 0 \tag{1.15}$$

It follows that the velocity of electromagnetic waves or any disturbance in the fields will propagate with velocity c, which is the ratio of the electromagnetic to the electrostatic units of electrical charge. The value of c as found experimentally is 2.99798×10^8 m/s, which agrees with the velocity of light in *vacuo*. This agreement of velocity establishes the identity of light with electromagnetic radiation.

In Maxwell's theory, an unobserved, all-pervasive medium, the *ether*, acts as a carrier of electromagnetic disturbances and c is the velocity of light measured by an observer, which is at rest relative to the ether. It was then conjectured that ether might provide the standard required to specify the Body Alpha and thus furnish the much-coveted link between the hitherto separated theories of mechanics and electromagnetism. If this were so, then one would expect the Maxwell's equations to hold in each and every inertial frame. To check this possibility, let us see whether Maxwell's equations are invariant under Galilean transformations, given by Eqs [1.1(a)]–[1.1(d)]. Consider Eq. [1.12(b)]

$$\nabla \times \mathbf{B} = \frac{1}{c}\frac{\partial \mathbf{E}}{\partial t} \tag{1.16}$$

whose x-component is

$$\frac{\partial B_z}{\partial y} - \frac{\partial B_y}{\partial z} = \frac{1}{c}\frac{\partial E_x}{\partial t}$$

This transforms into

$$\frac{\partial B'_z}{\partial y'} - \frac{\partial B'_y}{\partial z'} = \frac{1}{c}\left(\frac{\partial E'_x}{\partial t'} - \frac{\partial E'_x}{\partial x'}\right) \tag{1.17}$$

As there is no other equation in the set linking $\frac{\partial E'_x}{\partial t'}$ and $\frac{\partial E'_x}{\partial x'}$, so the transformed equation would never become

$$\frac{\partial B'_z}{\partial y'} - \frac{\partial B'_y}{\partial z'} = \frac{1}{c}\frac{\partial E'_x}{\partial t'} \tag{1.18}$$

which would have been the case if Maxwell's equations were invariant under Galilean transformation. Thus the form of the wave equation will not be preserved by the substitution of Eqs [1.1(a)]–[1.1(d)]. Presumably, the electromagnetic effects will not be the same if observed from two different frames moving with constant velocity with respect to each other. The velocity of propagation of plane waves will not retain its magnitude c.

The desired transformation must involve some change in the time coordinate such that $\frac{\partial t'}{\partial x}$ is non-zero. As will be obvious from Section 2.2, this transformation will abandon the concept of absolute time.

1.5 SEARCH FOR A UNIVERSAL FRAME OF REFERENCE

1.5.1 Michelson–Morley Experiment

The most famous attempt to find whether light similar to sound in air requires a medium ether for its propagation and secondly whether it is possible to detect the motion of earth relative to it, was the Michelson–Morley experiment[3] (1887), using the instrument Michelson had invented. This consisted in sending light signals simultaneously in two directions at right angles to each other and measuring the difference in the times they take to return to their starting point after they had been reflected at equal distances.

The experimental arrangement is shown in Figure 1.2. S is a source of light which sends light rays to fall on half-silvered mirror M. Roughly one-half of light is transmitted to M_1 from which it is reflected and the other half is reflected towards mirror M_2, which in turn reflects it back. The beams of light recombine after a second half-transmission at M and form an interference pattern in the field that the observer at O observes.

One of the mirrors was mounted on a micrometer screw and it is ensured that the motion of the movable mirror by only a fraction of a wavelength could be detected. The whole instrument was mounted on a slab of stone floating in mercury. Without disturbing the adjustment of the instrument, it could be rotated through 90° so that the two paths were interchanged. Were it so that one of the beams was at a

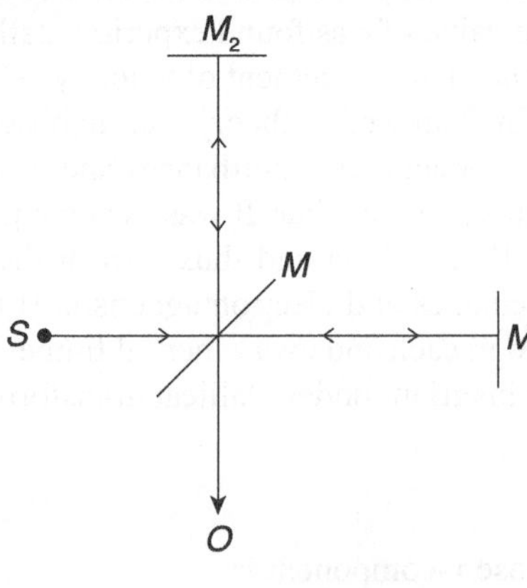

Figure 1.2 *Schematic diagram of Michelson–Morley experiment.*

disadvantage because of the direction of motion through the ether, this will cause a fringe shift in the interference pattern.

Suppose that c is the velocity of the light signals, **v** the velocity of the earth through ether, and earth is moving from east to west. If d is the length of the optical paths, then the times t_1 and t_2 along the two paths are

$$t_1 = \frac{d}{c-v} + \frac{d}{c+v} = \frac{2d}{c}\frac{1}{1-v^2/c^2} \tag{1.19}$$

$$t_2 = \frac{2d}{c}\frac{1}{\sqrt{1-v^2/c^2}} \tag{1.20}$$

When the arrangement is rotated through 90°, the new times are

$$\bar{t}_1 = \frac{2d}{c}\frac{1}{\sqrt{1-v^2/c^2}} \tag{1.21}$$

$$\bar{t}_2 = \frac{2d}{c}\frac{1}{1-v^2/c^2} \tag{1.22}$$

Now $$\Delta\bar{t} - \Delta t = (\bar{t}_2 - \bar{t}_1) - (t_2 - t_1)$$

$$= \frac{4d}{c}\frac{1}{\sqrt{1-v^2/c^2}}\left(\frac{1}{\sqrt{1-v^2/c^2}} - 1\right) \tag{1.23}$$

or the difference in optical path $$c(\Delta\bar{t} - \Delta t) = \frac{4d}{\sqrt{1-v^2/c^2}}\left(\frac{1}{\sqrt{1-v^2/c^2}} - 1\right) \tag{1.24}$$

This difference in path length produces a shift of n fringes given by

$$n = \frac{4d}{\lambda\sqrt{1-v^2/c^2}}\left(\frac{1}{\sqrt{1-v^2/c^2}} - 1\right)$$

$$\approx \frac{2d}{\lambda}\left(\frac{v^2}{c^2}\right) \tag{1.25}$$

when $v \ll c$.

The differences that the experiment is required to detect are of the second order in the ratio of $v = 3 \times 10^6$ cm/s, the velocity of earth in its orbit, and $c = 3 \times 10^{10}$ cm/s; and to be successful the experiment must be capable of detecting and showing changes of 1 part in 10.[8]

The experiment was repeated after an interval of 6 months, so as to eliminate the unlikely possibility that in a particular performance of the experiment, the earth might be at rest relative to ether at the time when observations are being taken. If this were so, then after 6 months the velocity of ether will be maximum.

No fringe shift corresponding to Eq. (1.25) was observed by Michelson and Morley and therefore no time difference. The estimated accuracy of the result was 10^3 cm/s, that is, the velocity of earth with respect to ether frame must be less than 10^3 cm/s. However, experiments of Miller[4] (1925) gave positive indication of a velocity of about 10 km/s, but a new analysis of Miller's experiments by Synge[5] (1952) and subsequent verification of the latter's theory by Ditchburn and Heavens[6] (1952) gave the verdict that Miller's data are not inconsistent with those of other observers.[7]

The surprising negative result of Michelson–Morley experiment was explained independently by Fitzgerald and Lorentz in 1892 by suggesting that the length of light path in the direction of earth's motion contracts automatically whereas the length of light path in the direction at right angles to the direction of earth's motion remains unaltered.

If we accept this ad hoc hypothesis, we can calculate the contraction factor easily. Taking the contraction factor as K and the time t_1 for the whole journey to and from SM M_1 gets modified to

$$t_1' = \frac{Kd}{c+v} + \frac{Kd}{c-v} = \frac{2Kd}{c}\frac{1}{1-v^2/c^2} \tag{1.26}$$

while t_2 is unaffected.
Substituting $t_1' = t_2$, we get

$$K = \sqrt{1 - v^2/c^2} \tag{1.27}$$

Thus, each unit of length of a moving body, in the direction of its motion, is contracted to $\sqrt{1 - v^2/c^2}$. Although this hypothesis did explain the Michelson–Morley experiment, its obvious weakness was that it was not capable of experimental verification. Since all scales placed against the interferometer arms for comparison would shrink in the same proportion, no change will ever be detected. But certain optical experiments were performed to detect this change of dimensions in glass, since it will be accompanied by strains and these in turn will manifest by phenomenon such as double refraction.[8,9]

The negative results of Michelson–Morley experiment are understandable if the postulate of ether is rejected. But it was Einstein[1] who provided the theory of relativity in 1905, which was a major reconstruction of the description of physical phenomena. He concluded that the velocity of light is always the same in all directions and is independent of the relative uniform motion of the observer, the transmitting medium, and the source.

1.5.1.1 Modern version of MM experiment

In the modern context, it is useful to view the Michelson–Morley (MM) experiment not as an interferometer experiment, but as a clock anisotropy experiment. Each arm of the interferometer can be thought of as a clock. Then one has to check whether the rate of a clock is independent of its orientation relative to its motion through the universe. Most modern incarnations of the MM experiment are clock anisotropy experiments. For example, MM experiments[10,11] using lasers compare two laser resonant cavities by beating their frequencies against each other, as one or both rotate relative to the universe. As a result of these laser experiments, the null prediction of the MM experiment was recovered.

1.5.2 Kennedy–Thorndyke Experiment

The Kennedy–Thorndyke[12] experiment, first carried out in 1932, is the modified form of Michelson–Morley experiment. The modification was to make one arm of the classical Michelson–Morley interferometer very short and the arms were not at right angles to each other. They used a spectacular technique to keep the apparatus temperature constant at 0.001°C, which gave them sufficient stability to record the observations during several seasons. In addition, they used photographs of their fringes. The apparatus was fixed to the earth and could only rotate with it.

In their own words, 'the principle on which this experiment is based is the simple proposition that if a beam of homogeneous light is split into two beams, which after traversing paths of different lengths are brought together again, then the relative phases will depend on the velocity of the apparatus (speed of the earth would cause changes in the travel times of the light rays), unless the frequency of light depends on the velocity in the way required by relativity.' The tests were done over a period of many months. As no significant fringe shift was found, they concluded that time dilation occurs according to the prediction of special relativity.

1.5.2.1 Modern re-enactments of the Kennedy–Thorndyke experiment[13–19]

Recent tests of special relativity are motivated by the developments in modern physics such as certain versions of string theory, which suggest that special relativity may not hold exactly and that violations might reveal themselves in tests with a certain level of precision.

An experiment by Brillet and Hall is one of the most accurate limits on any anisotropy in the round-trip speed of light in a laboratory. They measured the beat-frequency between a single-mode laser on a rotating table and a single-mode laser fixed to the earth to put a limit on such anisotropy of 3 parts in 10^{15}. Due to the construction of the rotating laser, this can also be interpreted as a limit on any anisotropy of space. This is a round-trip experiment because they used a Fabry–Perot etalon to determine the frequency of the rotating laser. Note that their limit on the round-trip anisotropy corresponds to a round-trip speed of less than 0.0000001 m/s; in terms of the more usual one-way anisotropy, it is 30 m/s.

Another experiment similar to Brillet and Hall was performed by Hills and Hall, but the lasers were fixed to the earth for better stability. No variations were found at the level of 2×10^{-13}. As they made observations over a year, this is not merely a limit on anisotropy but also on variations in different inertial frames.

The most precise test so far of special relativity, better than previous tests by a factor of three, has been performed by researchers at the universities of Konstanz and Dusseldorf in Germany. They devised a state-of-the-art version of the nineteenth century Michelson–Morley experiment, which first established that the speed of light is isotropic or the same in every direction. The speed of light does not depend on its direction of propagation to within 1.7 parts in 10^{15}, an accuracy about three times higher compared to the best previous experiment.

In their experiment, the Konstanz–Dusseldorf team made use of two devices known as optical cavities. Each cavity basically consists of two mirrors held at a constant distance. How long it takes a light beam to make a round trip between the mirrors is a direct measure of the speed of light perpendicular to the mirror surfaces. Each of the two cavities is oriented in a different direction. Rotating the set-up could measure light speed in a variety of directions. However, errors can accrue in this set-up through variations in the cavity length, caused by temperature effects and material ageing process. They employed cavities made from an ultra-pure sapphire crystal, which is virtually impervious to ageing effects, and operated it at the temperature of liquid helium at 4.2 K. Therefore, they reduced the cavity length significantly than what can be achieved at room temperature. Using advanced laser techniques for reading out the cavity round-trip time, the team obtained a new limit on possible violations of light propagation isotropy.

In another class of experiments, called Kennedy–Thorndyke (KT) measurements, one tests that *c* does not depend on the velocity of the laboratory. Since present MM precision is higher than the best KT precision, the Konstanz researchers aimed for a better KT test as a way of confirming, to a new level of accuracy, that *c* is independent of both the speed and the direction of laboratory. Using this technique, the team set up a standing wave of laser light in a cryogenic optical resonator or 'CORE,' in which the frequency of the standing wave can remain stable over very long periods. For 190 days, they compared its oscillation frequency with that of a standard iodine 'clock,' which is based on the radiation corresponding to the energy gap between two well-known electronic states in an iodine molecule. During this period, the earth traces out more than one-half of its orbit around the sun, altering the velocity of the laboratory by an amount equal to 60 km/s. If c were to vary with the speed, then the standing waves would fall out of tune with the cavity. The team found that frequency of the light in the CORE remained constant throughout the experiment, confirming that, over this velocity range, the speed of light is indeed independent of the velocity of the laboratory in which it is measured. This amounts to a verification of the special theory of relativity three times more accurate than the previous tests based on KT method. The researchers point out that the accuracy of their set-up was limited by the stability of the iodine clock. They believe that the recently developed optical clocks, which are around seven times more accurate than the best atomic clocks, will allow them to test special relativity even more stringently.

Summarily, on the basis of new generation of relativity tests, they confirmed the following:

1. The velocity independence of the speed of light to a three-fold higher accuracy,
2. Local position invariance,
3. Isotropy of space with accuracy being three times better than the previous experiment.

1.5.3 Trouton and Noble Experiment

Another experiment to measure the earth's motion through ether was performed by Trouton and Noble[20] (1902, 1903). A delicately suspended parallel plate condenser carried through the ether by earth's relative motion will experience a torque tending to rotate it into the longitudinal position, in which the plates are parallel to the direction of motion. The component of velocity parallel to the plates produces two equal and opposite currents, and the component of velocity perpendicular to the plates produces two currents almost neutralizing each other.

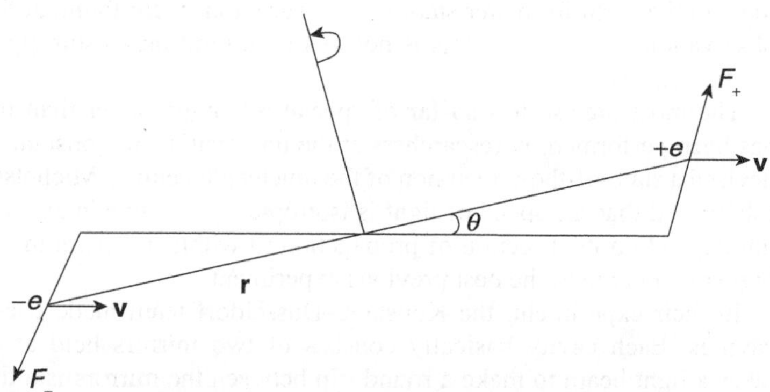

Figure 1.3 *Charges +e and −e move with velocity* **v**.

In order to compute the torque, it is convenient to imagine the plate charge e concentrated in two small spheres, whose separation is equal to the condenser plates (see Figure 1.3).

The translatory motion of the charge $+e$ due to the earth's motion corresponds to current Ids for which

$$I d\mathbf{s} = e\mathbf{v} \tag{1.28}$$

According to Biot–Savart law, this current element gives rise to a magnetic field **H** at the position of the negative charge given by

$$\mathbf{H} = \frac{I(d\mathbf{s} \times \mathbf{r})}{cr^3}$$

$$= \frac{e(\mathbf{v} \times \mathbf{r})}{cr^3} \tag{1.29}$$

where **r** is the vector drawn from the positive to the negative charge. Thus, the field **H** is normal to the plane of **v** and **r**.

The resulting force on the negative charge $-e$, which is also moving with velocity **v** is given by

$$\mathbf{F} = -e\frac{\mathbf{v}\times\mathbf{H}}{c} = -\frac{e^2}{c^2}\frac{(\mathbf{v}\times\mathbf{v}\times\mathbf{r})}{r^3}$$

$$|\mathbf{F}| = \frac{e^2v^2}{c^2}\frac{\sin\theta}{r^2} \tag{1.30}$$

An equal and opposite force acts upon the positive charge due to the field created by the motion of the negative charge. The torque parallel to the axis of rotation is

$$T = Fr\cos\theta = \frac{e^2}{r}\frac{v^2}{c^2}\sin\theta\cos\theta$$

$$= \frac{1}{2r}\frac{e^2v^2}{c^2}\sin 2\theta$$

$$= \frac{1}{2c^2}\frac{e^2v^2}{r}\sin 2\theta \tag{1.31}$$

This equation indicates that the effect is of the order of $(v/c)^2$ in comparison with the electrostatic interaction of the two charges. With $v = 30$ km/s, the velocity of motion of earth in its orbit; the effect should be easily measurable. The experiment performed by Trouton and Noble failed to give any indication of the torque. Thus, this experiment too, like the Michelson–Morley experiment, contradicts the theory of stationary ether.

1.5.4 Experiment with Ammonia Masers

An experiment for the velocity of ether wind and capable of accuracy far higher than the experiment of the Michelson–Morley type, was suggested by Møller[21] at the International Congress of Fundamental Constants in Turin in 1956. It consisted in comparing the frequencies of two masers; having their beams of NH_3 molecules, travel in opposite directions. A beam of excited ammonia molecules enters a resonant cavity and the beam is of sufficiently high intensity, so that the molecules in it maintain a monochromatic oscillation in the cavity due to the forced microwave transitions occurring in the molecules. The molecules enter the cavity with a certain velocity, so the frequency emitted in the transition, will depend on the direction of emission due to Doppler effect. The effect of an ether drift on the oscillation frequency of a beam-type laser is discussed as follows.

Let S' be the absolute system with time and space coordinates t', $\mathbf{x}'$, in which the ether is at rest. If S represents the laboratory system, the systems S' and S are connected by the Galilean transformation

$$t' = t$$

$$\mathbf{x}' = \mathbf{x} + \mathbf{v}t \tag{1.32}$$

where **v** is the absolute velocity of S. The forced transitions in the molecules are governed by Schrödinger equation, which is form-invariant under Galilean transformation Eq. (1.32), so the expression for transition probability has the same formula in both the systems.

Neglecting the effects of the magnetic field, we know that the transition probability is given by

$$p = -\frac{|\mu|^2}{\hbar^2}\left|\int_{t_0}^{t_0+L/u} E(t)\, e^{-2\pi i v_0 t}\, dt\right|^2 \tag{1.33}$$

where μ is the electric dipole matrix element for the transition; L, is the length of the cavity; and u is the velocity of the molecule, t_0 is the time of entrance of the molecule into the cavity, and $E(t)$ is the electric field at the position of the molecule.

If Maxwell's equations were also form-invariant under Galilean transformation, then in that case, the motion of maser through ether would not affect the frequency of emission. But this is certainly not the case (Section 1.4). Long back it was shown by Lorentz[22] that equations of electrodynamics attain the Maxwellian form, if we introduce the following transformations:

$$t_1 = t - \frac{\mathbf{v}\cdot\mathbf{x}}{c^2} = t' - \frac{\mathbf{v}\cdot\mathbf{x}}{c^2}$$
$$\mathbf{x}' = \mathbf{x} \tag{1.34}$$

where t_1 is the local time parameter. (The transformations Eq. (1.32) and Eq. (1.34) hold so far as we can neglect the second-order terms in v/c, otherwise these have to be replaced by exact Lorentz transformations.)

If any solution of Maxwell's equations satisfying the boundary conditions at the cavity walls, when the latter is at rest in S', is given by

$$\mathbf{E}'(\mathbf{x}', t') = \mathbf{f}(\mathbf{x}', t') \tag{1.35}$$
$$\mathbf{H}'(\mathbf{x}', t') = \mathbf{g}(\mathbf{x}', t') \tag{1.36}$$

then the solution of Maxwell's equations satisfying the boundary conditions when the cavity is at rest in S is

$$\mathbf{E}(\mathbf{x}, t_1) = \mathbf{f}(\mathbf{x}, t_1) \tag{1.37}$$
$$\mathbf{H}(\mathbf{x}, t_1) = \mathbf{g}(\mathbf{x}, t_1) \tag{1.38}$$

Consider the case when the maser is at rest in the ether. Let the electric field $\mathbf{E}'(\mathbf{x}', t')$ be harmonic in time, then

$$\mathbf{E}'(\mathbf{x}', t) = \mathbf{A}(\mathbf{x}')e^{2\pi i v t} \tag{1.39}$$

and

$$\mathbf{E}'(t) = \mathbf{A}(\mathbf{x}'_0 + \mathbf{u}t)\, e^{2\pi i v t} \tag{1.40}$$

where $\mathbf{x}'_o$ is the position of the molecule at the time t_0 and $\mathbf{u}$ is its velocity. Substituting Eq. (1.40) into Eq. (1.33), we get

$$p = -\frac{|\mu|^2}{\hbar^2}\left| \int_{t_0}^{t_0 + L/u} \mathbf{A}(\mathbf{x}'_0 + \mathbf{u}t)\, e^{2\pi i(v - v_0)t}\, dt \right|^2 \tag{1.41}$$

The power emitted by the beam has a maximum at the transition frequency $v = v_o$. Let us consider the same maser at rest in the system S with the absolute velocity $\mathbf{v}$. Then the electric field in S is given by

$$\mathbf{E}(\mathbf{x}, t) = \mathbf{A}(\mathbf{x})e^{2\pi i v t_1}$$
$$= \mathbf{A}(\mathbf{x})e^{2\pi i v t(t - \mathbf{v}\cdot\mathbf{x}/c^2)} \tag{1.42}$$

and the field at the place of the molecule is

$$\mathbf{E}(t) = \mathbf{A}(\mathbf{x}_0 + \mathbf{u}t)\, e^{2\pi i v[t - \mathbf{v}\cdot(\mathbf{x}_0 + \mathbf{u}t)/c^2]}$$
$$= \mathbf{A}(\mathbf{x}_0 + \mathbf{u}t)\, e^{-2\pi i v(\mathbf{v}\cdot\mathbf{x}_0/c^2)}\, e^{2\pi i v(1 - \mathbf{v}\cdot\mathbf{u}/c^2)t} \tag{1.43}$$

Introducing the expression into Eq. (1.33), we find that the power emitted by the beam has a maximum at a frequency ν_m given by

$$\nu_m\left(1-\frac{\mathbf{v}\cdot\mathbf{u}}{c^2}\right)=\nu_o \tag{1.44}$$

which, on neglecting higher order terms, becomes

$$\nu_m=\nu_o\left(1+\frac{\mathbf{v}\cdot\mathbf{u}}{c^2}\right) \tag{1.45}$$

Cedarholm et al.[23] carried out this experiment with two masers with opposite beam directions with the expected null result.

The arrangement was capable of detecting 1 part in 10^{12} and it was calculated that the velocity of ether wind must be smaller than 1/30 km/s, i.e., 45 times smaller than the upper limit claimed by Miller and about 1,000 times smaller than the velocity of earth.

The same experiment was repeated by Jaseja et al.[24] using He–Ne masers. The monochromaticity and stability of He–Ne masers and their sensitivity to fractional changes in length as small as 1×10^{13} suggested their use in testing ether drift or anisotropy in the velocity of light. Two masers were mounted perpendicular to each other on a massive shock-mounted platform that rotates periodically through 90° about its vertical axis, inside an acoustically isolated vault. First tests of the variation with rotation in the frequency of beat between the two masers showed that any frequency shift due to an ether drift is less than 1 part in $10.^{11}$

References

1. A. Einstein, *Ann. Phys.* (1905) 17, 891; English translation in *The Principle of Relativity; Zur Electrodynamik Bewegter Korper*, Dover Publication Inc., New York, 1958. Paperback.
2. F. E. Neumann, *Die Principien de Galilie–Newton'schen Theorie*, Leipzig, Teubner 1870.
3. A. A. Michelson and E. W. Morley, *The Relative Motion of the Earth and the Luminiferous Ether, Am. J. Sci.* (1887) 34, 333; *Phil. Mag.* (1887) 24, 449.
4. D. C. Miller, "Ether-drift experiments and the determination of the absolute motion of Earth", *Revs. Mod. Phys.* (1933) 5, 203–242; *Proc. Nat. Acad. Sci.* (1925) 9, 306.
5. J. L. Synge, *Effect of Acceleration in the Michelson–Morley Experiment, Sci. Proc. Roy. Dub. Soc.* (1952) 26, 45.
6. R. W. Ditchburn and O. S. Havens, *Relativistic Theory of a Rigid Body, Nature* (1952) 170, 705.
7. R. S. Shankland, S.W. McCuskey, F. C. Leone and G. Kuerti, *New Analysis of the Interferometric Observations of Dayton C. Miller; Rev. Mod. Phys.* (1955) 27, 167–178.
8. E. Dewan and M. Beran, "Note on the stress effects due to relativistic contraction", *Is Length Contraction Really Paradoxical; Am. J. Phys.* (1959) 27, 517–518; A. A. Evett and R. W. Wangness, *Am. J. Phys.* (1960) 28, 566.
9. P. J. Nawrocki, "Stress effects due to relativistic contraction", *Am. J. Phys.* (1962) 30, 771.
10. T. S. Jaseja, A. Javan, J. Murray and C. H. Townes, *Test of Special Relativity or of the Isotropy of Space by Use of Infrared Masers, Phys. Rev. A*(1964) 133, 1221.
11. A. Brillet and J. L. Hall, *An Optical Clock Based on a Single Trapped 199 Hg^+ ion, Phys. Rev. Lett.* (1979) 42, 549.
12. R. J. Kennedy and E. M. Thorndyke, "Experimental establishment of the relativity of time", *Phys. Rev.* (1932) 42, 400–418.

13. A. Brillet and J. L. Hall, "Improved laser test of the isotropy of space". *Phys. Rev. Lett.* (1979) 42, 549–552.

14. D. Hils and J. L. Hall, "Improved Kennedy-Thorndyke experiment to test special relativity" *Phys. Rev. Lett.* (1990) 64, 1697.

15. C. Braxaier, H. Muller, O. Pradl, J. Mlynek, A. Peters and S. Schiller, "Test of relativity using a cryogenic optical resonator", *Phys. Rev. Lett.* (2002) 88, 010401.

16. H. Muller, S. Herrmann, C. Braxmaier, S. Schiller and A. Peters, "Modern Michelson-Morley experiment using cryogenic optical resonators", *Phys. Rev. Lett.* (2003) 91, 020401.

17. H. Muller, C. Braxmaier, S. Herrmann, A. Peters, and C. Lammerzahl, "Electromagnetic cavities and Lorentz invariance violation", *Phys. Rev. D* (2003) 67, 056006.

18. Herrmann et al., "Test of the isotropy of the speed of light using a continuously rotating optical resonator", *Phys. Rev. Lett.* (2005) 95, 150401. Improved limits at a level of a few parts in 10^{-16}.

19. Chen et al., "Experimental test of the isotropy of a two-way light speed", *A. S. N. U. Peking* (1997) 33, 595. An experiment similar to Brillet and Hall, with a limit of 1×10^{-18} in the anisotropy of *c*.

20. F. T. Trouton and H. R. Noble, *Phil. Trans. Roy. Soc. Lond. A* (1903) 202, 165; *Proc. Roy. Soc. Lond.* (1903) 72, 132.

21. C. Møller, *Suppl. Nouvo. Cimento* (1957) 6, 381; *Proc. Roy. Soc. A* (1967) 270, 306.

22. H. A. Lorentz, *Enzykopadie der Mathematischaen Wissenschaften* (1903) Vol. 14.

23. C. J. Cedarholm, G. F. Bland, B. L. Havens and C. A. Townes, *Phy. Rev. Lett.* (1958) 1, 342. Also refer to C. H. Townes, *Advances in Quantum Electronics*, J. Singer ed., Columbia University Press, New York, 1961.

24. T. S. Jaseja, A. Javan, J. Murray and C. H. Townes, *B. A. P. S.* (1963) Series II, 8(4), 395.

Problems

1.1 Show that while momentum by itself is not invariant under Galilean transformation, its law of conservation is.

1.2 Show that the law of conservation of energy is invariant to Galilean transformation.

1.3 A train is moving uniformly and a particle is suspended from the roof of a carriage. The particle moves in a circular motion relative to the train. Use Galilean transformation to show that its path relative to the earth is a cycloid.

1.4 Calculate the fringe shift in Michelson–Morley experiment, given that effective length of each arm = 11 m, orbital velocity of earth = 3×10^6 cm/s, and $\lambda = 6000$ Å. (0.37 cm)

1.5 In Michelson interferometer, effective length of each arm = 11 m; v/c (= β) for earth's motion = 10^{-4}. To start with, one arm of the interferometer is parallel to earth's motion. If the apparatus is rotated through 90°, calculate the fringe shift expected. (No shift)

1.6 It is impossible to devise a mechanical experiment for detecting absolute uniform motion. Justify.

2

Lorentz Transformation and Its Kinematic Consequences. Intervals and Causality

2.1 EINSTEIN'S POSTULATES OF SPECIAL THEORY OF RELATIVITY. PRINCIPLE OF RELATIVITY AND CONSTANCY OF VELOCITY OF LIGHT

According to the theory of electromagnetic field as developed by Maxwell and Lorentz, the velocity of propagation of light or any electromagnetic disturbance in free space is a quantity c given by $c = 2.99792 \times 10^{10}$ cm/s. According to the basic correctness of Maxwell–Lorentz equations, Eqs [1.10(a)]–[1.10(d)], and classical kinematical laws, it follows that there is a unique frame called *ether frame*, in which Maxwell's equations are valid and light is propagated with velocity c (measured by an observer at rest relative to ether).

The laws of mechanics are invariant according to the principle of Galilean relativity, Eqs [1.10(a)]–[1.10(d)], but the equations of electrodynamics do not preserve their form when going from one system to another in uniform motion with respect to the former, Eq. (1.17). Thus the failure of Galilean relativity in case of electromagnetism sets the laws of electromagnetism apart from the laws of mechanics and in view of this controversy, the following situation emerges:

1. Principle of relativity exists for classical mechanics but not for electromagnetism. An ether frame acts as privileged inertial frame in electromagnetism.
2. Principle of relativity is valid both for mechanics and electromagnetism but Maxwell's equations need modification.
3. Principle of relativity is valid both for mechanics and electrodynamics but Newton's laws need modification.

The final conclusion regarding any of the aforementioned three issues rests with the outcome of experimental evidence. A number of experiments were conducted to decide between these hypotheses and these included observation of the aberration of star positions during the different parts of an year; measurement of the velocity of light in moving fluid, Fizeau's experiment[1]; and in addition a number of electrical, Trouton and Noble[2]; optical, Michelson–Morley[3], and microwave experiments[4] were performed to detect the ether wind due to motion of earth through it.

The aberration of starlight (small shift in apparent positions of distant stars during the year) finds a simple explanation in the motion of the earth in its orbit around the sun at a velocity of the order of 3×10^6 cm/s. It contradicts the hypothesis that the velocity of light is determined by the transmitting medium or that ether is carried along by earth. But the results of these experiments are judged, in general consensus, as negative. These negative results are understandable if the postulate of '*ether*' is rejected, but still the paradox of the failure of the law of addition of velocities remained. Further, the relative velocity between the two reference systems ought to show some effects on the wave propagation. Such an effect has never been found. The only logical but inevitable conclusion is that the laws of electromagnetic field hold in all inertial systems, with the same universal value of the velocity of light, viz., the above-mentioned alternative (3). Poincaré inferred that the motion referred to the supposed stationary ether cannot be observed and stated the principle of relativity which says that the laws of physics are the same for two observers moving with constant velocity relative to each other.

Lorentz discovered the required transformations to replace Eqs (1.1), which leave Maxwell's equations invariant. Lorentz arrived at space–time transformations, called Lorentz transformations, and it was in this connection that he introduced for the sake of mathematical convenience, a local time t' for a moving system of reference,[5] and its introduction left Maxwell's equations in charge-free space invariant. The general invariance of Maxwell's equations was, however, proven by Poincaré,[6] who introduced the correct transformations for charge and current densities. But no serious attempt was made to relate the space and time coordinates to physical observation. Einstein[7] in his epoch-making paper of 1905 not only independently deduced all the essential results deduced by Poincaré and Lorentz, but also gave these transformations a physical meaning. He was probably unaware of the work of Lorentz and Poincaré since his own first paper is devoid of any references. Later he called the transformation equations after Lorentz. He showed that Newton's laws are not invariant under Lorentz transformations and thus a mechanical experiment could lead to the discovery of a privileged reference system, if Newton's laws are assumed correct. Einstein modified Newton's laws to make them invariant under Lorentz transformations and thus did away completely with the privileged frame of reference both for electromagnetic and mechanical experiments. The combination of Poincaré principle of relativity with the finiteness of the velocity of propagation of interaction, called the special theory of relativity, was put forth by Einstein in 1905. It expresses the mutual equivalence of all inertial systems moving relative to one another and its postulates may be stated as follows:

1. **The postulate of relativity:** It is impossible to trace any distinction between any two reference systems which are in uniform relative motion to each other, by any physical measurement.
2. **The postulate of the constancy of velocity of light:** The velocity of light is always the same in all directions and is independent of the relative motion of the observer, the medium, and the source.

The mechanics based on these two postulates is called relativistic and goes over into Newtonian or classical mechanics for the case when $v << c$ or in the limit $c \to \infty$.

The concept of absolute time is in contradiction with the principle of relativity, since the law of addition of velocities based on the concept of absolute time will predict different velocities of propagation in different inertial systems and thus be in contradiction to the postulate of the constancy of velocity of light. Moreover, the experiment of Michelson–Morley demonstrates the lack of dependence of the velocity of light on the direction of propagation. This sequel is in contradiction to the predictions of classical mechanics, based on the concept of absolute time. For a more detailed information on the foundations of the special theory of relativity, the readers may refer to treatise by Mϕller[8] (1952). A detailed discussion of the logical and philosophical foundations of the special theory of relativity is given by Grünbaum.[9]

According to postulate (2), Maxwell's equations work in an identical form in different inertial frames in vacuo. We derive the wave equation from the set of Maxwell's equations in the Gaussian units.

$$\nabla \times \mathbf{E} = -\frac{1}{c}\frac{\partial \mathbf{B}}{\partial t} \tag{2.1}$$

$$\nabla \cdot \mathbf{B} = 0 \tag{2.2}$$

$$\nabla \times \mathbf{H} = \frac{4\pi \mathbf{J}}{c} + \frac{1}{c}\frac{\partial \mathbf{D}}{\partial t} \tag{2.3}$$

$$\nabla \cdot \mathbf{D} = 4\pi\rho \tag{2.4}$$

where the symbols have their usual meaning. Taking the curl of Eq. (2.1), we get

$$\nabla \times (\nabla \times \mathbf{E}) = -\frac{1}{c}\nabla \times \frac{\partial \mathbf{B}}{\partial t} = -\frac{1}{c}\frac{\partial}{\partial t}(\nabla \times \mathbf{B})$$

or

$$\nabla(\nabla \cdot \mathbf{E}) - \nabla^2 \mathbf{E} = -\frac{\partial}{c\,\partial t}(\nabla \times \mathbf{B}) \tag{2.5}$$

Substituting the value of $(\nabla \times \mathbf{B})$ from Eq. (2.3) into Eq. (2.5), we obtain

$$\nabla(\nabla \cdot \mathbf{E}) - \nabla^2 \mathbf{E} = -\frac{\mu}{c}\frac{\partial}{\partial t}\left(\frac{4\pi \mathbf{J}}{c} + \frac{1}{c}\frac{\partial \mathbf{D}}{\partial t}\right)$$

where we used the relation $\mathbf{B} = \mu\mathbf{H}$. In a region where there is no charge, $\rho = 0$, and using the relations $\mathbf{D} = \varepsilon\mathbf{E}$ and $\mathbf{J} = \sigma\mathbf{E}$, the equation becomes

$$\nabla^2 \mathbf{E} - \frac{4\pi\mu\sigma}{c^2}\frac{\partial \mathbf{E}}{\partial t} - \frac{\mu\varepsilon}{c^2}\frac{\partial^2 \mathbf{E}}{\partial t^2} = 0 \tag{2.6}$$

However, for an isotropic dielectric medium, $\sigma = 0$ and Eq. (2.6) can be written as

$$\left(\nabla^2 - \frac{\mu\varepsilon}{c^2}\frac{\partial^2}{\partial t^2}\right)\mathbf{E} = 0 \tag{2.7}$$

This equation is equivalent to three scalar equations satisfied by the three components of $\mathbf{E}$. For the x-component of $\mathbf{E}$, the equation becomes

$$\left(\nabla^2 - \frac{\mu\varepsilon}{c^2}\frac{\partial^2}{\partial t^2}\right)E_x = 0 \tag{2.8}$$

Equation (2.8) has non-zero solutions, implying thereby that electromagnetic fields can exist in the absence of charges. Such electromagnetic fields are called electromagnetic waves. An equation of type (2.8) is called a wave equation and is typically written as

$$\left(\nabla^2 - \frac{1}{v^2}\frac{\partial^2}{\partial t^2}\right)E_x = 0 \tag{2.9}$$

where $v = c/\sqrt{\mu\varepsilon} = c/n$ is the phase velocity of the wave and n is the refractive index of the medium in which the wave propagates. As one will notice, the velocity of light in a medium depends only on the values of two constants ε (the permittivity) and μ (the permeability). In vacuo, these constants are ε_0 (the permittivity of free space) and μ_0 (the permeability of free space). The process of swapping from one frame to another does not affect the value of the velocity of light. Hence postulate 2 can be derived from postulate 1.

2.1.1 Invariance of the Velocity of Light

Several classic experiments have been performed to check whether the speed of light depends on the velocity of the emitter. If the speed of light is given by $c + kv$, where v is the velocity of the emitter, k is a parameter to be measured or bounded, then the orbits of binary stars would appear to have an anomalous eccentricity, which is not explainable by usual Newtonian gravity. However, this result is complicated by another phenomenon, called extinction, i.e., losing the memory of the speed of light. Since at optical wavelengths, the light will be absorbed and re-emitted by the intervening space, there will be an ambiguity in the results obtained. However, at x-ray wavelengths, the path length of extinction is tens of kilometres, so a nearby x-ray binary source in our galaxy is ideal for checking the validity of the proposal. Using data on pulsed 70 kev x-ray binary systems—Hers-1, Cen x–3, and SMC x–1—Brecher[10] obtained a bound $|k| < 2 \times 10^{-9}$ for typical orbital velocities $v/c \sim 10^{-3}$.

In another experiment conducted at CERN in 1964 by Alväger et al.,[11] ultra-relativistic particles were used as the source of light. Neutral pions were produced by the collision of 20 GeV protons on stationary nucleons in the proton synchrotron. With energies greater than 6 GeV, the pions had $v/c \geq 0.99975$. Photons resulting from the decay $\pi^0 \rightarrow \gamma_1 + \gamma_2$ were collimated and timed over 30-m long path length. Because the protons in the synchrotron were pulsed, the speed of the photons could be measured from the arrival times of their pulses as a function of the varying location of the detector along the flight path. The result for the speed was $2.9977 \pm 0.0004 \times 10^8$ m/s, in agreement with the laboratory value. This experiment thus sets a bound $|k| < 10^{-4}$ for $v \approx c$.

2.2 LORENTZ TRANSFORMATIONS[12]. AXIOMATIC DERIVATION

An event is anything that can either happen in space–time or can be imagined, such as the emission of light, collision between two particles, or a particle just passing by. It is defined by the three coordinates of space and the time of its occurrence, and the purpose here is to derive the general relations between the time and space coordinates of an event as observed from inertial frames in relative motion.

Consider two reference systems S and S' moving relative to each other with constant velocity $\mathbf{v}$, Figure 2.1. The coordinates are so chosen that the OX and $O'X'$ axes coincide at $t = 0$, while the OY and OZ axes are parallel to $O'Y'$ and $O'Z'$. If the particular event is specified by the coordinates (x, y, z, t) in S, then the same event is represented by the coordinates (x', y', z', t') in S'. For finding the requisite transformations between the unprimed and primed quantities, the following general requirements should be built in them:

1. The postulate of the equivalence of both the inertial systems S and S' require that the direct (for going from the unprimed to the primed) and reverse (for going from the primed to the unprimed) transformations should be symmetrical with respect to both systems, that is, one should be

derivable from the other by changing the sign of v and converting unprimed into primed quantities and vice versa.

2. The requisite transformations should convert the finite points of one system into the finite points of the other, that is, if (x, y, z, t) are finite, then (x', y', z', t') should also be finite.
3. In the limit when $v \to 0$, the Lorentz transformations should reduce to an identity transformation, that is, $x' = x$, $y' = y$, $z' = z$, and $t' = t$.
4. In the limit when $c \to \infty$, the Lorentz transformations should reduce to Galilean transformations, Eqs (1.1), viz.,

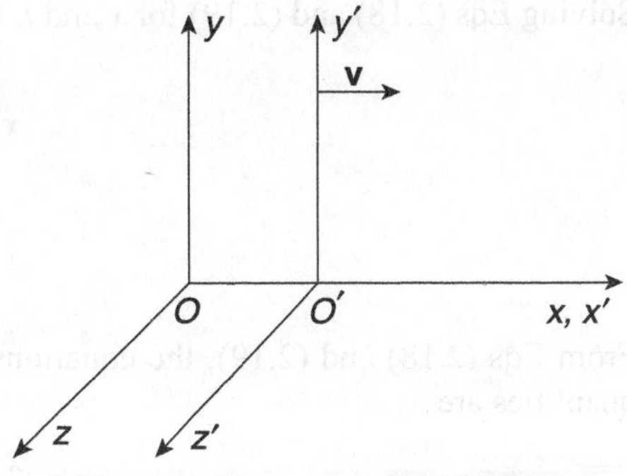

Figure 2.1 *The primed system is moving with velocity v relative to the unprimed system.*

$$x' = x - vt \tag{2.10}$$

$$y' = y \tag{2.11}$$

$$z' = z \tag{2.12}$$

$$t' = t \tag{2.13}$$

5. The law of addition of velocities obtained from these transformations should leave the velocity of light *in vacuo* invariant, i.e., $c' = c$.

From the aforementioned conditions (1–5), we learn that the possible form of the transformation is greatly restricted. The transformation functions cannot be quadratic or of any degree higher than the first, since then inversion will lead to irrationality. However, a linear-fractional transformation may be inverted and still retain the same form, such as

$$x' = \frac{ax + b}{cx + d} \tag{2.14}$$

which gives

$$x = \frac{b - dx'}{cx' - a} \tag{2.15}$$

But this function for $x' = a/c$ becomes infinite and thus leads to violation of condition (2). Consequently, a linear transformation function is the only acceptable one. From Figure 2.1, since the relative motion is along the x-axis, we get

$$y' = y \tag{2.16}$$

$$z' = z \tag{2.17}$$

The linear transformations of x and t are

$$x' = \lambda x + \mu t \tag{2.18}$$

$$t' = \psi x + \phi t \tag{2.19}$$

There are no constants on the right-hand sides of Eqs (2.18) and (2.19). This is because the origin initially is at $x = 0$, $t = 0$ in one system and at $x' = 0$ and $t' = 0$ in the other.

Let us apply Eq. (2.18) to the origin of the primed system $x' = 0$. Since this point O' moves with velocity v relative to O, we get $x = vt$. Thus,

$$\lambda v + \mu = 0 \tag{2.20}$$

Solving Eqs (2.18) and (2.19) for x and t, we get

$$x = \frac{\phi x' - \mu t'}{\lambda\phi - \mu\psi} \tag{2.21}$$

$$t = \frac{\lambda t' - \psi x'}{\lambda\phi - \mu\psi} \tag{2.22}$$

From Eqs (2.18) and (2.19), the equations for the inverse transformations from unprimed to primed quantities are

$$x = \lambda x' - \mu t' \tag{2.23}$$

$$t = -\psi x' + \phi t' \tag{2.24}$$

The coefficients μ and ψ, which interrelate the coordinate and time, must change sign when $v \to -v$. Comparing Eqs (2.23) and (2.21), we obtain

$$\lambda = \frac{\phi}{\lambda\phi - \mu\psi} \tag{2.25}$$

$$-\mu = \frac{-\mu}{\lambda\phi - \mu\psi} \tag{2.26}$$

From Eq. (2.26), we get

$$\lambda\phi - \mu\psi = 1 \tag{2.27}$$

and so from Eq. (2.25)

$$\lambda = \phi \tag{2.28}$$

This is the relation obtained by comparing the direct equations with the inverse equations. Let us employ the condition (5).
Dividing Eq. (2.18) by Eq. (2.19), we get

$$\frac{x'}{t'} = \frac{\lambda\frac{x}{t} + \mu}{\psi\frac{x}{t} + \phi} \tag{2.29}$$

If x is the point occupied by a light signal emitted from the origin of the unprimed system at an initial instant of time $t = 0$, then $x/t = c$. Hence from Eq. (2.29),

$$c = \frac{\lambda c + \mu}{\psi c + \phi} \tag{2.30}$$

Substituting the relations (2.20) and (2.28) into Eq. (2.30), we get

$$\begin{aligned} \psi c^2 + \lambda c &= \lambda c + \mu \\ &= \lambda c - \lambda v \end{aligned}$$

Hence,

$$\psi = -\lambda \frac{v}{c^2} \tag{2.31}$$

Further, substituting Eqs (2.31), (2.20), and (2.28) into (2.27), we get

$$\lambda^2 - \lambda^2 \frac{v^2}{c^2} = 1$$

or

$$\lambda = \pm \frac{1}{\sqrt{1 - v^2/c^2}} \tag{2.32}$$

According to condition (3), Eq. (2.19) becomes $t' = t$ when $v \to 0$, hence we must select the positive sign of the square root. Thus, we get λ, μ, ψ, and ϕ from Eqs (2.32), (2.20), (2.31), and (2.28), respectively

$$\lambda = \frac{1}{\sqrt{1 - v^2/c^2}} \tag{2.33}$$

$$\mu = -\frac{v}{\sqrt{1 - v^2/c^2}} \tag{2.34}$$

$$\psi = -\frac{v/c^2}{\sqrt{1 - v^2/c^2}} \tag{2.35}$$

and

$$\phi = \frac{v}{\sqrt{1 - v^2/c^2}} \tag{2.36}$$

Substituting these results into Eqs (2.18) and (2.19), we get

$$x' = \frac{x - vt}{\sqrt{1 - v^2/c^2}} \tag{2.37}$$

$$t' = \frac{t - vx/c^2}{\sqrt{1 - v^2/c^2}} \tag{2.38}$$

The complete Lorentz transformations are the following:

$$x' = \gamma(x - vt) \qquad [2.39(a)]$$

$$y' = y \qquad [2.39(b)]$$

$$z' = z \qquad [2.39(c)]$$

$$t' = \gamma(t - vx/c^2) \qquad [2.39(d)]$$

where $\gamma = (1 - v^2/c^2)^{-1/2}$, and it is called the Lorentz contraction factor.
The inverse Lorentz transformations are given by

$$x = \gamma(x' + vt') \qquad [2.40(a)]$$

$$y = y' \qquad [2.40(b)]$$

$$z = z' \qquad [2.40(c)]$$

$$t = \gamma(t' + vx'/c^2) \qquad [2.40(d)]$$

We consider two sets of (x, y, z, t) and (x', y', z', t') coordinates, which are related to each other by Eqs (2.39) and (2.40) as *Lorentz frames of reference.*

Example 2.1 (i) Show that the interval s, defined by

$$s^2 \equiv c^2t^2 - x^2 - y^2 - z^2$$

is invariant under Lorentz transformation.
(ii) Prove that the differential interval between two events

$$ds^2 \equiv c^2dt^2 - dx^2 - dy^2 - dz^2$$

is invariant under Lorentz transformation.

Solution. (i) The invariance of interval implies that

$$c^2t^2 - x^2 - y^2 - z^2 = c^2t'^2 - x'^2 - y'^2 - z'^2$$

where (ct, x, y, z) and (ct', x', y', z') are the coordinates of the same event in the unprimed and primed systems, respectively.

Since

$$y' = y;\ z' = z,$$

we get

$$c^2t^2 - x^2 = c^2t'^2 - x'^2$$

Now

$$c^2t'^2 - x'^2 = \frac{c^2\,(t - vx/c^2)^2 - (x - vt)^2}{(1 - v^2/c^2)}$$

$$= \frac{c^2t^2\,(1 - v^2/c^2) - x^2\,(1 - v^2/c^2)}{(1 - v^2/c^2)}$$

$$= c^2t^2 - x^2$$

Thus the interval is invariant under Lorentz transformation.
(ii) Taking differentials of the coordinates and the time in Eq. (2.39), we find that

$$dx' = \gamma(dx - vdt)$$

$$dy' = dy$$

$$dz' = z$$

$$dt' = \gamma\left(dt - \frac{vdx}{c^2}\right)$$

where $\gamma = (1 - v^2/c^2)^{-1/2}$. Thus, (dx, dy, dz, dt) transform in the same way as (x, y, z, t) and hence ds^2 is invariant under Lorentz transformations.

Example 2.2 Assuming the invariance of the equations of wave front in two inertial systems, which are moving with uniform velocity relative to each other, that is

$$c^2t^2 - x^2 - y^2 - z^2 = c^2t'^2 - x'^2 - y'^2 - z'^2$$

derive the Lorentz transformation.

Solution. Since the relative motion is in the x-direction, we assume that

$$y' = y;\ z' = z$$

hold and so we have

$$c^2t^2 - x^2 = c^2t'^2 - x'^2 \tag{i}$$

The identical transformations $x' = x$ and $t' = t$ are trivial, and x, t had to transform separately. Since at any instant $x' = 0$ corresponds to $x = vt$, we write $x' = \gamma(x - vt)$ and the inverse transformation

$$x = \gamma'(x' + vt')$$

or

$$x/\gamma' = \gamma x - v\gamma t + vt'$$

which gives

$$t' = \gamma\left[t - \frac{x}{v}\left(1 - \frac{1}{\gamma\gamma'}\right)\right] \tag{ii}$$

From Eq. (i),

$$\begin{aligned} c^2t'^2 - x'^2 &= \gamma^2t^2(c^2 - v^2) + 2\gamma^2xt\left[v - \frac{c^2}{v}\left(1 - \frac{1}{\gamma\gamma'}\right)\right] - \gamma^2x^2\left[1 - \frac{c^2}{v^2}\left(1 - \frac{1}{\gamma\gamma'}\right)^2\right] \\ &= c^2t^2 - x^2 \end{aligned}$$

which will be satisfied provided

$$\gamma = \pm\frac{1}{\sqrt{1 - v^2/c^2}} \tag{ii}$$

and

$$\gamma' = \pm\frac{1}{\sqrt{1 - v^2/c^2}}$$

But, $x' = x$ for $v = 0$, we take the positive sign of the square root. We thus obtain the Lorentz transformations

$$\begin{aligned} x' &= \gamma(x - vt) \\ y' &= y \\ z' &= z \\ t' &= \gamma\left(t - \frac{vx}{c^2}\right) \end{aligned}$$

where $\gamma = (1 - v^2/c^2)^{-1/2}$.

Example 2.3 Show that d'Alembertian operator

$$\Box^2 \equiv \frac{\partial^2}{\partial x^2} + \frac{\partial^2}{\partial y^2} + \frac{\partial^2}{\partial z^2} - \frac{1}{c^2}\frac{\partial^2}{\partial t^2}$$

is invariant under Lorentz transformation.

Solution. From Eqs (2.39), we get

$$\frac{\partial}{\partial x'} = \gamma\left(\frac{\partial}{\partial x} + \frac{v}{c^2}\frac{\partial}{\partial t}\right)$$

$$\frac{\partial}{\partial y'} = \frac{\partial}{\partial y}; \frac{\partial}{\partial z'} = \frac{\partial}{\partial z}$$

$$\frac{\partial}{\partial t'} = \gamma\left(v\frac{\partial}{\partial x} + \frac{\partial}{dt}\right)$$

Therefore,

$$\frac{\partial^2}{\partial x'^2} = \gamma^2\left(\frac{\partial^2}{\partial x^2} + \frac{2v}{c^2}\frac{\partial^2}{\partial x \partial t} + \frac{v^2}{c^4}\frac{\partial^2}{\partial t^2}\right)$$

$$\frac{\partial^2}{\partial y'^2} = \frac{\partial^2}{\partial y^2}; \frac{\partial^2}{\partial z'^2} = \frac{\partial^2}{\partial z^2}$$

$$\frac{\partial^2}{\partial t'^2} = \gamma^2\left(v^2\frac{\partial^2}{\partial x^2} + 2v\frac{\partial^2}{\partial x \partial t} + \frac{\partial^2}{\partial t^2}\right)$$

So,

$$\frac{\partial^2}{\partial x'^2} + \frac{\partial^2}{\partial y'^2} + \frac{\partial^2}{\partial z'^2} - \frac{1}{c^2}\frac{\partial^2}{\partial t'^2} = \frac{\partial^2}{\partial x^2} + \frac{\partial^2}{\partial y^2} + \frac{\partial^2}{\partial z^2} - \frac{1}{c^2}\frac{\partial^2}{\partial t^2}$$

Example 2.4 Show that two successive Lorentz transformations with velocity parameters γ_1 and γ_2 are equivalent to a single Lorentz transformation of parameter

$$\gamma = \gamma_1\gamma_2(1 + v_1v_2/c^2)$$

where v_1 and v_2 are two velocities with respect to which the two Lorentz transformations are applied in the same direction.

Solution. From Lorentz transformations

$$x_1 = \gamma_1(x - v_1t)$$

$$t_1 = \gamma_1(t - v_1x/c^2)$$

$$x_2 = \gamma_2(x_1 - v_2t_1)$$

so $$x_2 = \gamma_1\gamma_2(1 + v_1v_2/c^2)(x - vt)$$

where $$v = \frac{(v_1 + v_2)}{(1 + v_1v_2/c^2)}.$$

It can be shown that successive Lorentz transformations, in general, do not commute; they commute only when successive velocities are parallel. In this case, as the velocities are in the same direction, the resultant transformation does not depend on the order of application of the successive Lorentz transformations. Further,

$$1 - \frac{v^2}{c^2} = 1 - \frac{(v_1 + v_2)^2}{c^2\left(1 + \frac{v_1v_2}{c^2}\right)^2} = \frac{(1 - v_1^2/c^2)(1 - v_2^2/c^2)}{\left(1 + \frac{v_1v_2}{c^2}\right)^2}$$

or $$\gamma(v) = \gamma(v_1)\,\gamma(v_2)$$

or

$$\gamma(v) = \gamma_1\gamma_2(1 + v_1v_2/c^2)$$

Example 2.5 Derive the equations for Lorentz transformation for an arbitrary direction of the velocity **v** relative to a coordinate system.

Solution. From Eq. (2.39), we have the formulae for special Lorentz transformation

$$x' = \gamma(x - vt)$$

$$y' = y$$

$$z' = z$$

$$t' = \gamma\left(t - \frac{vx}{c^2}\right)$$

Here

$$x = \frac{\mathbf{r}\cdot\mathbf{v}}{v}$$

$$x' = \frac{\mathbf{r}'\cdot\mathbf{v}}{v}$$

The component perpendicular to the velocity is

$$\mathbf{r} - \frac{\mathbf{v}(\mathbf{r}\cdot\mathbf{v})}{v^2} = \mathbf{r}' - \frac{\mathbf{v}(\mathbf{r}'\cdot\mathbf{v})}{v^2} \quad \text{(i)}$$

$$\frac{\mathbf{r}'\cdot\mathbf{v}}{v} = \gamma\left(\frac{\mathbf{r}.\mathbf{v}}{v} - vt\right) \quad \text{(ii)}$$

Multiplying Eq. (ii) by $\mathbf{v}/v$ and adding Eqs (i) and (ii), we get

$$\mathbf{r}' = \mathbf{r} - \frac{\mathbf{v}(\mathbf{r}\cdot\mathbf{v})}{v^2} + \gamma\left(\frac{\mathbf{v}(\mathbf{r}\cdot\mathbf{v})}{v^2} - \mathbf{v}t\right)$$

$$t' = \gamma\left(t - \frac{\mathbf{r}\cdot\mathbf{v}}{c^2}\right)$$

These are the requisite Lorentz transformations for an arbitrary direction of **v**.

2.3 KINEMATIC CONSEQUENCES OF LORENTZ TRANSFORMATION

2.3.1 Lorentz–Fitzgerald Contraction

Consider a rod at rest in the system S placed parallel to the x-axis, and let its length as measured by an observer in S be $l = x_2 - x_1$, where x_2 and x_1 are the coordinates of the two ends of the rod in the system. If another observer in the system S' measures it at any instant, say t', what will be the measured length of this rod?

From Eq. (2.40), we get

$$x_1 = \frac{x_1' + vt'}{\sqrt{1 - v^2/c^2}} \tag{2.41}$$

and

$$x_2 = \frac{x_2' + vt'}{\sqrt{1 - v^2/c^2}} \tag{2.42}$$

Let the length in the system S' be $l' = (x_2' - x_1')$, then

$$l = l'(1 - v^2/c^2)^{-1/2} \qquad [2.43(a)]$$

or

$$l' = l(1 - v^2/c^2)^{1/2} \qquad [2.43(b)]$$

The *proper length* of a rod is its length in a reference system in which it is at rest. Thus, the proper length of a rod is the greatest and to any other observer, which is moving with velocity v, the rod appears to be contracted in the ratio $[1 - v^2/c^2]^{1/2}:1$. Furthermore, this effect is reciprocal.

This result differs from the contraction hypothesis postulated by Lorentz and Fitzgerald to explain the negative results of Michelson–Morley experiment. This is because Eq. (2.43b) gives a symmetrical relation between two measuring sticks in relative motion, while hypothesis required a change in length for a single rod depending on its actual velocity through a real fixed ether.

Since

$$y = y'$$

and

$$z = z'$$

the transverse dimensions do not change because of motion. The volume V' of a body is given by

$$V' = V(1 - v^2/c^2)^{1/2} \tag{2.44}$$

where V is the proper volume of the body.

Although the act of seeing involves a single observer, the observation of Lorentz–Fitzgerald contraction requires the collaboration of at least two observers. It is shown by Weinstein[13] that the length seen by a single observer is not the usual contraction and indeed, under certain circumstances, one sees a body considerably lengthened rather than contracted. The calculations are as follows.

A rod moving along the x-direction with velocity **v** relative to the laboratory is viewed by the observer O in the laboratory, (see Figure 2.2). ρ_1 and ρ_2 are the distances of the trailing and leading edges, respectively, of the rod at the time of observation t. Due to the finite velocity of light, the signals from both the edges must originate at different times, so as to be observed at the same time t by the observer, viz., at $(t - \rho_1/c)$ from the trailing edge and at $(t - \rho_2/c)$ from the leading edge. Since the leading and trailing edges of the rod are observed at different times, the experiment of one observer viewing the rod is not the same as the experiment resulting in the usually defined length.

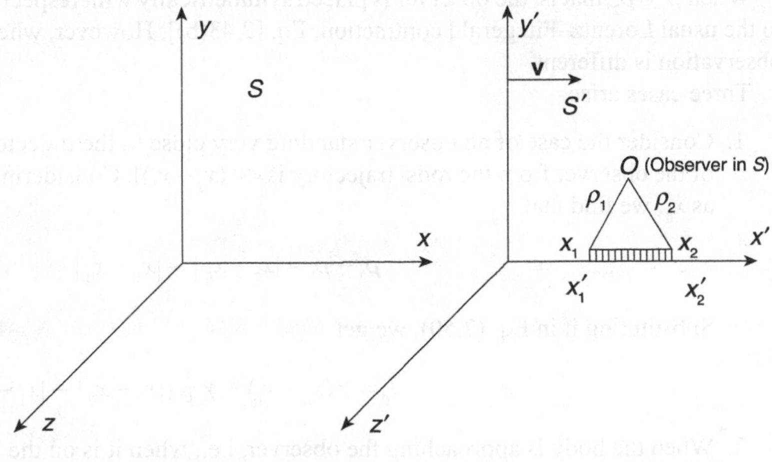

Figure 2.2 *Derivation of single observer results.*

The observer O will compare the apparent positions of the ends of the rod with markings on his x-coordinate axis and see a viewed length $l_v \equiv x_2 - x_1$.

Let t_0 be the time when the observer sees points x_1 and x_2 in his coordinate system. He received a light pulse, which left x_1 at a time

$$t_1 = t_0 - (\rho_1/c) \tag{2.45}$$

Similarly, the pulse from x_2 started at

$$t_2 = t_0 - (\rho_2/c) \tag{2.46}$$

In the primed system, the point corresponding to x_1, t_1 is from Eq. (2.39)

$$x_1' = \gamma(x_1 - vt_1) \tag{2.47}$$

Also,

$$x_2' = \gamma(x_2 - vt_2) \tag{2.48}$$

where $\gamma = (1 - v^2/c^2)^{-1/2}$.

Calling $x_2' - x_1' \equiv l_0$, the rest length of the rod, we get

$$l_0 = x_2' - x_1' = \gamma(x_2 - x_1 + vt_1 - vt_2) \tag{2.49}$$

Substituting Eqs (2.45) and (2.46) into Eq. (2.49)

$$l_0 = \gamma(x_2 - x_1) + \gamma\frac{v}{c}(\rho_2 - \rho_1) \tag{2.50}$$

When $\rho_1 = \rho_2$, that is, the observer is placed symmetrically with respect to the rod, Eq. (2.50) reduces to the usual Lorentz–Fitzgerald contraction, Eq. [2.43(b)]. However, when $\rho_2 \neq \rho_1$, the outcome of the observation is different.

Three cases arise:

1. Consider the case of an observer standing very close to the trajectory of a rod [i.e., the distance of the observer from the rods' trajectory is $<< (x_2 - x_1)$]. Considering the position of the observer as x_0, we find that

$$\rho_2 - \rho_1 \approx |x_2 - x_0| - |x_1 - x_0| \tag{2.51}$$

Substituting it in Eq. (2.50), we get

$$l_0 = \gamma (x_2 - x_1) + \gamma \tfrac{v}{c} [|x_2 - x_0| - |x_1 - x_0|] \tag{2.52}$$

2. When the body is approaching the observer, i.e., when it is on the left of the observer in Figure 2.1, then

$$\rho_2 - \rho_1 = |x_2 - x_0| - |x_1 - x_0| = x_2 - x_1 \tag{2.53}$$

Hence,

$$l_0 = \gamma (x_2 - x_1) + \gamma \tfrac{v}{c} (x_2 - x_1) = \gamma (x_2 - x_1) \left(1 + \tfrac{v}{c}\right) \tag{2.54}$$

Since $(x_2 - x_1)$ is the length seen by the observer, setting $(x_2 - x_1) \equiv l_v$, we find

$$l_v = l_0 \frac{\sqrt{1 - v/c}}{\sqrt{1 + v/c}} \tag{2.55}$$

The observer sees a contraction much more severe than the Lorentz–Fitzgerald contraction. Instead of the usual second-order effect, it is the first-order effect.

3. When the body recedes from the observer, he sees the reverse effect. In this case

$$\rho_1 - \rho_2 = |x_1 - x_0| - |x_2 - x_0| = x_1 - x_2 \tag{2.56}$$

and

$$l_v = l_0 \frac{\sqrt{1 + v/c}}{\sqrt{1 - v/c}} \tag{2.57}$$

These conclusions are in no way a contradiction of the special theory of relativity, but merely point out the deviation of the observation from the usual Lorentz–Fitzgerald contraction, when a single observer views a moving rod.

2.3.1.1 Test of length contraction

At present there are no direct tests of length contraction, as measuring the length of a moving object to the precision required has not been feasible. However, there is a demonstration that it occurs. A current-carrying wire is observed to be electrically neutral in its rest frame, and a nearby charged particle in that frame is unaffected by the current. A nearby charged particle that is moving parallel to the wire, however, is subject to a magnetic force that is related to the speed relative to the wire. If one considers the situation in the rest frame of a charge moving with the drift velocity of the electrons in the wire, the force is purely electrostatic due to different length contractions of the positive and negative charges in the wire (the former are fixed relative to the wire, while the latter are mobile with drift velocities of a few millimetres per second). This approach gives the correct quantitative value of the magnetic force in the wire frame. This is discussed in more detail in Purcel's *Electricity and Magnetism*.[14] It is indeed remarkable that the relativistic effects for such a diminutive velocity explain the enormous magnetic effects observed.

2.3.2 Time Dilation or Apparent Slowing of Moving Clocks

Suppose an observer from system (S') sends a light signal from the point (x', 0, 0) at t_1' and at a subsequent time t_2'. We here find how the interval ($t_2' - t_1'$) appears to an observer in the system S. From Eq. [2.40(d)], the corresponding times are

$$t_1 = \gamma\left(t_1' + \frac{vx'}{c^2}\right)$$

$$t_2 = \gamma\left(t_2' + \frac{vx'}{c^2}\right)$$

hence

$$(t_2 - t_1) = \gamma(t_2' - t_1') \tag{2.58}$$

which shows that the interval is increased in the ratio $1 : \sqrt{1 - v^2/c^2}$. Designating

$$t_2 - t_1 = \tau \qquad t_2' - t_1' = \tau'$$

from Eq. (2.58), we get

$$\tau = \tau'/\sqrt{1 - v^2/c^2} \tag{2.59}$$

The time read by a clock moving with a given object is called the *proper* time for this object. From Eq. (2.59), we see that the proper time of a moving object is always less than the corresponding interval in the rest frame. In other words, the greater the relative motion between the clock and observer, the more slowly a clock will be found to run. This kinematical effect of relativity is called *time dilation*.[15]

The effect of time dilation becomes important for high-energy particles such as electrons and mesons. If the particle is unstable, for example, $\pi^{\pm}$, π^{0}, $\mu^{\pm}$, etc., the lifetime is always considerably greater than τ_0, the lifetime measured in decay at rest.

A classic experiment performed by Rossi and Hall[16] showed that the lifetime of μ-mesons was prolonged by the standard factor $\gamma = 1/\sqrt{1 - v^2/c^2}$. Muons are created in the upper atmosphere when cosmic protons collide with nuclei of air producing pions, which decay to muons. With a rest half-life of 2.2×10^{-6} s, a muon travelling near the speed of light should travel only two-third of a kilometre on average before decaying to an electron or positron and two neutrinos. Yet, muons are the primary component of cosmic radiation detected at sea level. But with time dilation and a typical speed of $v/c \sim 0.994$, their lives in the rest frame of earth get prolonged by a factor of nine, enabling them to reach the earth.

Rossi and Hall measured the distribution of muons as a function of altitude and also measured their energies, thus confirming the time dilation formula.

In an experiment[17]performed in 1966 at CERN, muons produced by collisions at one of the targets in the accelerator were deflected by magnets so that these could move in circular paths in a 'storage ring'. Their speeds were 99.7 per cent of c, and the observed 12-fold increase in their lifetime agreed with the prediction with 2 per cent accuracy.

The apparent slowing down of clocks was demonstrated by using Mössbauer effect[18] in an ultracentrifuge to show that an observer moving at right angles to the source (where Doppler effect is excluded) will observe the frequency of monochromatic gamma-rays shifted[19] in accordance with the prediction of special relativity.

2.3.2.1 *Improved test of time dilation in special relativity*[20]

An improved test of time dilation in special relativity has been performed using laser spectroscopy on fast ions at the heavy ion storage-ring TSR in Heidelberg. In a collaboration with the university of Mainz, the Doppler-shifted frequencies of a two-level transition in $^7Li^+$ ions at $v = 0.064c$ have been measured in forward and backward directions to an accuracy of $\Delta v/v = 1 \times 10^{-9}$ using collinear saturation spectroscopy. The result is the modern version of the classical Ives–Stilwell experiment,[1] which confirms the relativistic Doppler formula and sets a new limit of 2.2×10^{-7} for deviations from the time dilation factor $\gamma = 1/\sqrt{1 - v^2/c^2}$.

2.3.2.2 *The clock hypothesis*

The clock hypothesis states that the tick rate of a clock measured in an inertial frame depends only upon its velocity relative to that frame and is independent of its acceleration or higher derivatives. The experiment of Bailey et al.[21, 22] stored muons in a magnetic storage ring and measured their lifetime. While being stored in the ring, they were subjected to a proper acceleration of approximately $10^{18}g$ ($1g = 9.8$ m/s^2). The observed agreement between the lifetime of the stored muons with the same energy moving inertially confirms the clock hypothesis for acceleration of that magnitude. Sherwin[23] discusses some Mössbauer experiments which show that the rate of a clock is independent of acceleration $\sim 10^{16}g$ and depends only upon velocity.

2.3.3 Relativity of Simultaneity

If two events are simultaneous in the system S but occur at different places at $x = x_1$ and $x = x_2$, then these will not be observed to be simultaneous in the system S'. If $t_1 = t_2$, then from Eq. [2.39(d)], we have

$$t_1' = \gamma\left(t_1 - \frac{vx_1}{c^2}\right) \tag{2.60}$$

and

$$t_2' = \gamma\left(t_2 - \frac{vx_2}{c^2}\right) \tag{2.61}$$

Thus, $t_1' \neq t_2'$ unless $x_1 = x_2$.

This theorem is reciprocal in character. A discussion on relativity of simultaneity in a single Galilean frame had been presented by Liebowitz[24] and Grünbaum.[25]

2.3.4 Causality and Maximum Signal Velocity

Let us consider how causality, i.e., the cause-and-effect relationship, puts an upper limit on the signal velocity. Earlier, we inferred that two events that are simultaneous in S will be simultaneous in S' only if both occur at the same place in system S. One can even state that the order in which two events occur can be different in different inertial frames. This is evident from the equation

$$t_2' - t_1' = \gamma\left((t_2 - t_1) - \frac{v}{c^2}(x_2 - x_1)\right) \tag{2.62}$$

since the signs of $(t_2' - t_1')$ and $(t_2 - t_1)$ can be different. This, however, violates the principle of causality, which implies that the cause has to necessarily precede the effect in time sequence.

Let us imagine two events A and B and let event A occur when the origins O and O' coincide and give rise to event B which occurs at x_B at a later time t_B in S and at x_B' and t_B' in S'. Causality imposes the condition that event B cannot occur earlier than A and as such $t_B' > 0$. Event A may be throwing a ball towards a window and event B the breaking of the window pane. The signal is propagated by the ball from A at a velocity v_1 in frame S.

Therefore,

$$\begin{aligned} t_B' &= \gamma\left(t - \frac{v}{c^2}x_B\right) \\ &= \gamma\left(t_B - \frac{v}{c^2}v_1 t_B\right) \\ &= \gamma\left(t_B\left(1 - \frac{vv_1}{c^2}\right)\right) \end{aligned} \tag{2.63}$$

As γ is a positive factor and by definition t_B is also positive; however, if v and v_1 have opposite signs, t_B' will always be positive and event B will occur later than event A. But if v and v_1 have the same sign, then it is imperative for the following inequality to hold:

$$vv_1 << c^2 \tag{2.64}$$

Assuming that v and v_1 are both positive and noting that v can be large but never greater than c, the above inequality will hold only when

$$v_1 << c \tag{2.65}$$

Thus a signal can never be propagated at a velocity larger than the velocity of light; otherwise the principle of causality will be violated since the effect will precede the cause.

2.3.5 Addition of Velocities

Let us derive the formulae relating the velocity of a material particle in one reference system to its velocity in another reference system. Let the components of velocity of the particle in system S be

$$u_x = \frac{dx}{dt} \tag{2.66}$$

and these in S' be

$$u_x' = \frac{dx'}{dt'} \tag{2.67}$$

From Eqs [2.39(a)]–[2.39(d)], we have

$$dx' = \gamma (dx - vdt) \qquad [2.68(a)]$$

$$dy' = dy \qquad [2.68(b)]$$

$$dz' = dz \qquad [2.68(c)]$$

$$dt' = \gamma \left(dt - \frac{vdx}{c^2}\right) \qquad [2.68(d)]$$

Dividing the first three equations by the fourth, we get

$$\frac{dx'}{dt'} = \frac{dx - vdt}{dt - vdx/c^2} \qquad [2.69(a)]$$

$$\frac{dy'}{dt'} = \frac{dy}{\gamma(dt - vdx/c^2)} \qquad [2.69(b)]$$

$$\frac{dz'}{dt'} = \frac{dz}{\gamma(dt - vdx/c^2)} \qquad [2.69(c)]$$

and dividing the numerator and denominator of the right-hand sides of these equations by dt and using Eq. (2.66), we get

$$u_x' = \frac{u_x - v}{1 - vu_x/c^2} \qquad [2.70(a)]$$

$$u_y' = \frac{u_y}{\gamma(1 - vu_x/c^2)} \qquad [2.70(b)]$$

$$u_z' = \frac{u_z}{\gamma(1 - vu_x/c^2)} \qquad [2.70(c)]$$

These expressions give Lorentz transformation for velocity components in frame S to those in frame S'. These essentially provide formulae for adding velocity components of the particle and the velocity of the frame S' w.r.t. S.

The inverse transformations are

$$u_x = \frac{u_x' + v}{1 + vu_x/c^2} \qquad [2.71(a)]$$

$$u_y = \frac{u_y'}{\gamma(1 + vu_x'/c^2)} \qquad [2.71(b)]$$

$$u_z = \frac{u_z'}{\gamma(1 + vu_x'/c^2)} \qquad [2.71(c)]$$

In the limiting case, when $v/c << 1$ or $c \to \infty$, Eqs [2.71(a)]–[2.71(c)] go over into

$$u_x = u'_x + v \quad [2.72(a)]$$

$$u_y = u'_y \quad [2.72(b)]$$

$$u_z = u'_z \quad [2.72(c)]$$

which are the classical results.

Let us consider the special case of motion of the particle parallel to the x-axis, then

$$u_x = u; \quad u_y = u_z = 0$$

Hence, from Eqs [2.70(a)]–[2.70(c)], we get

$$u'_y = u'_z = 0, \; u'_x = u' \text{ (say)}$$

and from Eq. [2.71(a)]

$$u = \frac{u' + v}{1 + u'v/c^2}$$

or

$$u' = \frac{u - v}{1 - uv/c^2} \quad (2.73)$$

When $u = c$, then

$$u' = \frac{c - v}{1 - cv/c^2} = c \quad (2.74)$$

Thus when a particle is moving with velocity c, as observed by an observer in S, its velocity as observed in S' is also c. The velocity of the frame S' has no effect on the velocity of light. This is a verification of the fact that Lorentz transformation is consistent with the postulate of the constancy of velocity of light and hence in accordance with the results of Michelson–Morley experiment.

Although the velocity of light c appears to be an upper limit for velocities, but velocities greater than c are known. For a medium with refractive index less than unity such as plasma, the wave velocity is greater than c. But signals can be given only by pulses or groups of waves that are propagated with the group velocity, which is always less than the velocity of light. So any wave which transfers any physical quantity, such as energy, field, or momentum, cannot have velocity greater than c.

2.3.6 The Fizeau Effect

The Einstein velocity addition theorem gives the simplest explanation of index of refraction of moving bodies (Fizeau effect). Fizeau employed an interferometer to measure the velocity of light in liquids flowing in a pipe, both in the direction of and opposed to the propagation of light.

A homogeneous isotropic medium of refractive index n moves with velocity v along the x-axis relative to an observer in S. The velocity of light with respect to the frame in which the medium is at rest is $u' = c/n$. If v is the velocity with respect to an observer at rest in S, then from Eq. (2.73) we get

$$u = \frac{c/n + v}{1 + v/(nc)} \approx \frac{c}{n} + v\left(1 - \frac{1}{n^2}\right) \quad (2.75)$$

to the first order in v/c.

2.3.7 Aberration of Light

We choose our coordinate axes in such a way that the velocity of the particle at a given moment lies in the XY-plane. The trajectory of the particle subtends angles θ and θ' with OX and $O'X'$ in systems S and S', respectively, and $\mathbf{u}$ and $\mathbf{u}'$ are its velocities in S and S', then

$$u_x = u\cos\theta;\ u_y = u\sin\theta \tag{2.76}$$

$$u'_x = u'\cos\theta';\ u'_y = u'\sin\theta' \tag{2.77}$$

From Eq. (2.71), we get

$$u_y/u_x = \frac{u'_y\sqrt{1 - v^2/c^2}}{u'_x + v} \tag{2.78}$$

Let us apply Eq. (2.78) to the case of a light beam and compute the deviation of it in transforming to a new coordinate system. This phenomenon of the change in the direction of velocity of light is known as the aberration of light.

So in this case, $u = u' = c$. Hence from Eq. (2.78), we get

$$\tan\theta = \frac{\sqrt{1 - v^2/c^2}\,\sin\theta'}{v/c + \cos\theta'} \tag{2.79}$$

Also, from Eq. (2.79), we get

$$\sin\theta = \frac{\sqrt{1 - v^2/c^2}\,\sin\theta'}{1 + v\cos\theta'/c} \tag{2.80}$$

$$\cos\theta = \frac{\cos\theta' + v/c}{1 + v\cos\theta'/c} \tag{2.81}$$

When $v << c$, from Eq. (2.80) we get

$$\sin\theta - \sin\theta' = -v\cos\theta'\sin\theta/c$$

Introducing $\Delta\theta = \theta' - \theta$, then

$$\Delta\theta \approx \frac{v}{c}\sin\theta$$

correct to terms of order v/c. This is the elementary formula for aberration of light.

Example 2.6 Show that the four-dimensional volume element $dx\ dy\ dz\ dt$ is invariant under Lorentz transformation.

Solution. From Eqs [2.43(a)] and (2.59), we have

$$dx = dx'\left(1 - \frac{v^2}{c^2}\right)^{\frac{1}{2}}$$

$$dt = dt'\left(1 - \frac{v^2}{c^2}\right)^{-\frac{1}{2}}$$

Now

$$dx\,dy\,dz\,dt = J\,dx'\,dy'\,dz'\,dt'$$

where J is the Jacobian of the transformation and is given by

$$J = \frac{\partial(x, y, z, t)}{\partial(x', y', z', t')} = \begin{vmatrix} \frac{\partial x}{\partial x'} & \frac{\partial y}{\partial x'} & \frac{\partial z}{\partial x'} & \frac{\partial t}{\partial x'} \\ \cdots & \cdots & \cdots & \cdots \\ \cdots & \cdots & \cdots & \cdots \\ \frac{\partial x}{\partial t'} & \cdots & \cdots & \frac{\partial t}{\partial t'} \end{vmatrix}$$

It can be shown that $J = 1$. Hence the result follows.

2.4 GEOMETRIC INTERPRETATION OF LORENTZ TRANSFORMATION

Lorentz transformation is a linear transformation in four-dimensional space. That leaves the differential interval defined by

$$ds^2 = c^2dt^2 - dx^2 - dy^2 - dz^2 \tag{2.82}$$

invariant. It was introduced by *Minkowski* that ct be treated as fourth coordinate. Designating

$$x_1 = x \qquad [2.83(a)]$$

$$x_2 = y \qquad [2.83(b)]$$

$$x_3 = z \qquad [2.83(c)]$$

$$x_4 = ict \qquad [2.83(d)]$$

and $$dS = ids \tag{2.84}$$

we get $$dS^2 = (dx_1)^2 + (dx_2)^2 + (dx_3)^2 + (dx_4)^2 \tag{2.85}$$

Since the special Lorentz transformation leaves dx_2 and dx_3 unaffected, we will only consider the invariance of the two-dimensional interval

$$dS^2 = (dx_1)^2 + (dx_4)^2 \tag{2.86}$$

This interval is invariant to translations of the origin and also to rotations of the coordinate axes in the $x_1 - x_4$ plane.

From Figure 2.3, we see that

$$x'_1 = x_1 \cos\theta + x_4 \sin\theta$$

$$x'_4 = -x_1 \sin\theta + x_4 \cos\theta \tag{2.87}$$

Substituting $\theta = i\phi$ and transforming back to x and t, we get

$$x' = x\cosh\phi - ct\sinh\phi \quad [2.88(a)]$$

$$ct' = -x\sinh\phi + ct\cosh\phi \quad [2.88(b)]$$

Equations (2.88) are identical with Lorentz transformations (Eqs [2.89(a)] and [2.91(d)]) provided

$$\sinh\phi = \frac{\beta}{\sqrt{1-\beta^2}}$$

$$\cosh\phi = \frac{1}{\sqrt{1-\beta^2}}$$

$$\tanh\phi = \beta$$

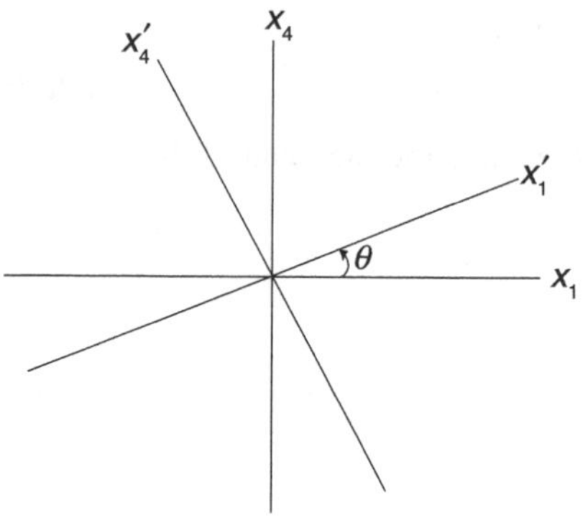

Figure 2.3 *Rotation of the x_1, x_4 axes through an angle θ into x_1', x_4' axes.*

where $\beta = v/c$.

Equations (2.88) show that Lorentz transformation corresponds to a rotation in four-dimensional space consisting of three dimensions of ordinary space plus a fourth imaginary dimension proportional to time. The Lorentz transformation is thus an orthogonal transformation in Minkowski space. Equation (2.59), which gives the resultant of two velocities, simply corresponds to the addition formula for $\tanh\phi$

$$\tanh(\phi_1 + \phi_2) = \frac{\tanh\phi_1 + \tanh\phi_2}{1 + \tanh\phi_1 \tanh\phi_2}$$

Let us investigate the real geometrical representation of Lorentz transformation in the real four-dimensional space of ct, x, y, z. In order to permit representation in a plane, we will plot only in x, ct plane. Such a two-dimensional representation permits the confirmation of all the properties of Lorentz transformation and thus gives a convenient geometrical interpretation to physical events. The trajectory of an event so plotted as a function of space and time is called a *world line* and the diagram itself is called the *Minkowski diagram* (See Figure 2.4). In Euclidean geometry, intervals are measured by the sums of the squares of the coordinate intervals and in this real space (x, ct) a substitute for distance measurement is obtained by noting that the family of hyperbolas

$$(x_1)^2 - (x_0)^2 = (x_1')^2 - (x_0')^2 = \text{constant}$$

lays out a convenient net which permits comparison of the various quantities. Such a net is shown by dotted lines in Figure 2.5.

In the *Minkowski* or x, ct diagram, each material point will be associated with a world line, which enables us to read the position of the point along the x-axis as a function of time. An event has objective or absolute reality and two events that coincide at (ct, x, y, z) will do so, irrespective of the frame of reference from which this is viewed. The spatial coincidence or temporal coincidence have only relative meaning and vary from one system to another.

The world line of a ray of light in vacuo is the line $x_1 = x_0$ at 45° to each of x_1, x_0 axes, Figure 2.4. The axes x_1 and x_0 transform into x_1' and x_0', Eqs [2.88(a)] and [2.88(b)], but the *world line* light ray (dotted line) is unchanged, since in each of the frames, the light ray has the same velocity c. The world line of a particle, which is at rest relative to a system S', is parallel to the x'_0-axis (L). Since a material point cannot travel with velocity greater than the velocity of light c, it follows that the world lines of all material particles are directed more steeply than the light line.

Let us further draw the *Minkowski diagram* to confirm the contraction of length and dilation of time, when observed from a moving system. Let a ruler have the length $AO = 1$, as observed by an observer at rest in S. The world lines of its ends are $0x_4$ and AA'. The spatial distance relative to S' between any pair of simultaneous events, that belong to these world lines, is given by intersects that are parallel to x'_1 axis, so to distance OA'. As is obvious from Figure 2.5, the length OA is smaller than the intercept of the member of hyperbola family. Hence to the observer in S' (moving with velocity v relative to S) the rod appears shorter than the unit length.

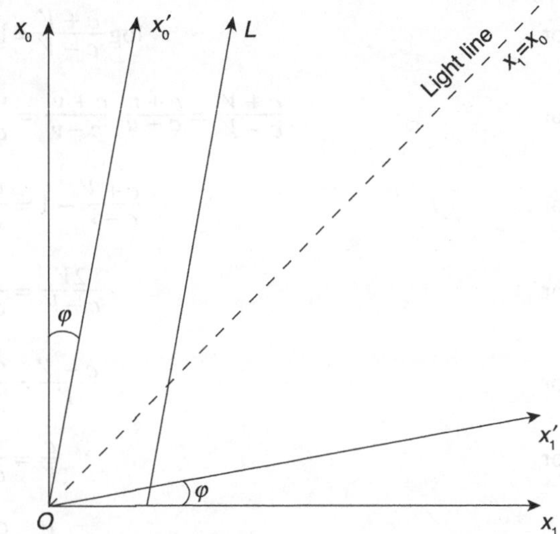

Figure 2.4 *Minkowski diagram showing transformed Lorentz axes.*

Imagine a clock at rest at the origin of S'. The world line is thus Ox'_4. The interval of a unit time in the primed system is given by OP'. The time distance between both the ends of this interval, according to S, are obtained by drawing parallels to the x_1-axis. A parallel drawn from P' intersects the Ox_0 axis at P. Now OP is greater than OQ (where $Q'Q$ is drawn parallel to Ox'_1 axis). Hence, according to the observer in S, more than a unit time has elapsed and accordingly the clock in S' appears slow.

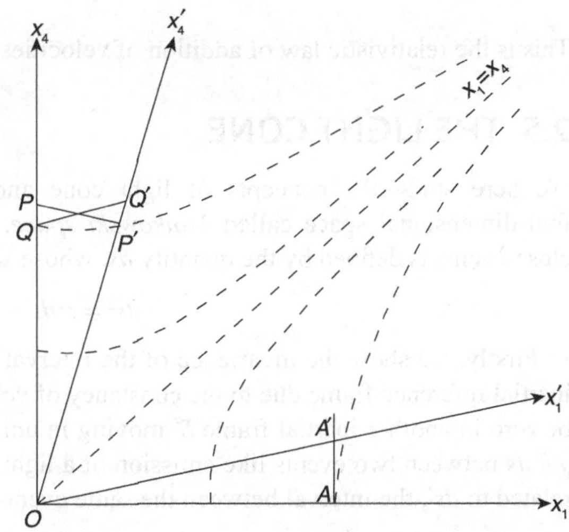

Figure 2.5 *Lorentz contraction and time dilation in Minkowski diagram.*

Example 2.7 If u and v are two velocities in the same direction and V their resultant velocity given by

$$\tanh^{-1}\frac{V}{c} = \tanh^{-1}\frac{u}{c} + \tanh^{-1}\frac{v}{c}$$

then deuce the law of composition of velocities from this result.

Solution. Given that

$$\tanh^{-1}\frac{V}{c} = \tanh^{-1}\frac{u}{c} + \tanh^{-1}\frac{v}{c}$$

Expressing it as

$$\frac{1}{2}\log\frac{c+V}{c-V} = \frac{1}{2}\log\frac{c+u}{c-u} + \frac{1}{2}\log\frac{c+v}{c-v}$$

or
$$\log \frac{c+V}{c-V} = \log \frac{c+u}{c-u}\frac{c+v}{c-v}$$

or
$$\frac{c+V}{c-V} = \frac{c+u}{c-u}\frac{c+v}{c-v} = \frac{c^2+(u+v)c+uv}{c^2-(u+v)c+uv}$$

or
$$\frac{c+V}{c-V} - 1 = \frac{c^2+(u+v)c+uv}{c^2-(u+v)c+uv} - 1$$

or
$$\frac{2V}{c-V} = \frac{2\,(u+v)c}{c^2-(u+v)c+uv}$$

or
$$\frac{c-V}{V} = \frac{c^2-(u+v)c+uv}{(u+v)c}$$

or
$$\frac{c}{V} = \frac{c^2+uv}{c(u+v)} + \frac{uv}{c(u+v)} = \frac{c^2+uv}{c(u+v)}$$

or
$$\frac{V}{c} = \frac{c(u+v)}{c^2+uv}$$

which gives
$$V = \frac{u+v}{1+\frac{uv}{c^2}}$$

This is the relativistic law of addition of velocities.

2.5 THE LIGHT CONE

We here study the concepts of light cone and interval between two events in an imaginary four-dimensional space called *Minkowski space*. The infinitesimal interval between two infinitely close events is defined by the quantity ds, whose square is given by

$$ds^2 = c^2dt^2 - dx^2 - dy^2 - dz^2 \tag{2.92}$$

Firstly, we show the invariance of the interval between two events. Equation (2.92) holds in any inertial reference frame due to the constancy of velocity of light. If it is zero in inertial frame S, it will be zero in another inertial frame S' moving in uniform motion w.r.t. S. Thus it follows that the interval ds between two events like emission of a light signal and its arrival at another point in S' must be related to ds', the interval between the same events in S' by the equation

$$ds = ads' \tag{2.93}$$

According to the principle of equivalence of all inertial frames, we have

$$ds' = ads \tag{2.94}$$

Substituting Eq. (2.94) into Eq. (2.93) gives,

$$ds = a^2ds$$

$$a^2 = 1$$

or
$$a = \pm 1 \tag{2.95}$$

However, the sign of the interval between two events must be the same. Therefore, choosing the positive sign, Equation (2.93) becomes

$$ds = ds' \tag{2.96}$$

Thus the interval between two events is invariant and this invariance is a direct consequence of the equivalence of all inertial frames. Since the invariant metric of Minkowski space is indefinite, the quadratic form $c^2dt^2 - \sum_i (dx_i)^2$ may take either sign. Due to the indefiniteness of the metric, the geometrical properties of Minkowski space are different from those of Euclidean space.

Let us consider an event whose world point is 0, the origin of the x, ct coordinates in Figure 2.6. We will investigate the relations of other events on the given event 0. Since the velocity of light is an upper limit on all velocities, the space–time domain is divisible into three regions

1. Light-like or $ds^2 = 0$
2. Time-like or $ds^2 > 0$
3. Space-like or $ds^2 < 0$

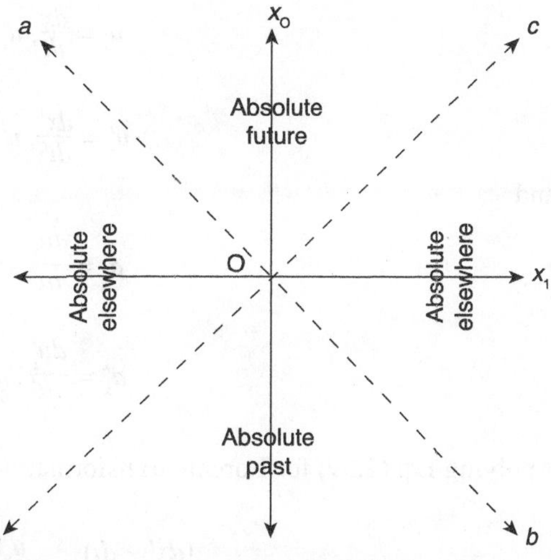

Figure 2.6 *Minkowski diagram of cone dividing space time into past, future, and elsewhere relative to an event 0.*

Let us consider these regions one by one.

1. Light-like or $ds^2 = 0$: The dotted lines $x = ct$ and $x = -ct$ represent respectively the propagation of two light signals in opposite directions through the event 0. Had we considered all the space coordinates x, y, z instead of just x, we would have got the cone $x^2 + y^2 + z^2 = c^2t^2$ instead of the light (or null) line. Light signals emitted at $t = 0$ from the origin travel out along the dotted lines and all these events are on the light cone.
2. Time like or $ds^2 > 0$: Let us consider events whose world points lie within the region *a0c*. It is evident that for all points of this region, the equation is positive, or the interval between any event in *a0c* and event *0* is time-like. Since t is positive, all the events in this region occur after the event *0*. Since the time-like character of an event is independent of the reference system, it is impossible to find a reference system in which any of the events in *a0c* occurred before the event *0*. Hence the region *a0c* is the absolute future relative to *0*.

   Analogously, the region *b0d* is absolute past relative to 0, that is, it is impossible to choose a reference system in which any event of the region occurred after the event 0.
3. Space-like or $ds^2 < 0$: Further, let us consider regions *d0a* and *b0c*. The interval between any event in these regions and event *0* is space-like, so any event from these regions and *0* occur at different points in space in any reference system. Thus the regions *d0a* and *b0c* are absolutely remote relative to the event *0* and are outside the light cone. For any event, one could find reference systems in these regions, in which event *0* could occur either earlier or later and one such system in which it occurred simultaneously with *0*.

2.6 TRANSFORMATION OF VELOCITY AND ACCELERATION

Consider a point in a primed system moving with a velocity **u** relative to the unprimed coordinate system and let us find the transformation between the components of velocity as measured by the observers in the two respective frames of reference. Defining

$$u_x = \frac{dx}{dt},\ u_y = \frac{dy}{dt},\ u_z = \frac{dz}{dt} \tag{2.97}$$

$$u'_x = \frac{dx'}{dt'},\ u'_y = \frac{dy'}{dt'},\ u'_z = \frac{dz'}{dt'} \tag{2.98}$$

and

$$a_x = \frac{du_x}{dt},\ a_y = \frac{du_y}{dt},\ a_z = \frac{du_z}{dt} \tag{2.99}$$

$$a'_x = \frac{du'_x}{dt'},\ a'_y = \frac{du'_y}{dt'},\ a'_z = \frac{du'_z}{dt'} \tag{2.100}$$

Applying Eq. (2.39) for Lorentz transformation, we get

$$u'_x = \frac{(dx' - dt)}{(dt'/dt)} = \frac{u_x - v}{1 - vu_x/c^2} \tag{2.101(a)}$$

$$u'_y = \frac{(dy'/dt)}{(dt'/dt)} = \frac{dy/dt}{\gamma(dt - vdx/c^2)/dt} = \frac{u_y}{\gamma(1 - vu_x/c^2)} \tag{2.101(b)}$$

$$u'_z = \frac{u_z}{\gamma(1 - vu_x/c^2)} \tag{2.101(c)}$$

where $\gamma = (1 - v^2/c^2)^{-1/2}$. In the limit $c \rightarrow \infty$, the above equations become

$$u'_x = u_x - v \tag{2.102(a)}$$

$$u'_y = u_y \tag{2.102(b)}$$

$$u'_z = u_z \tag{2.102(c)}$$

which are the Newtonian forms.

The resultant velocity is given by

$$u'^2 = \frac{1}{\left(1 - \frac{\mathbf{u \cdot v}}{c^2}\right)^2}\left\{(u_x - v)^2 + \frac{1}{\gamma^2}\left(u_y^2 + u_z^2\right)\right\}$$

$$= \frac{1}{\left(1 - \frac{\mathbf{u}\cdot\mathbf{v}}{c^2}\right)^2} \left\{ u_x^2 - 2\mathbf{u}\cdot\mathbf{v} + v^2 + \left(1 - \frac{v^2}{c^2}\right)(u_y^2 + u_z^2) \right\}$$

$$= \frac{1}{\left(1 - \frac{\mathbf{u}\cdot\mathbf{v}}{c^2}\right)^2} \left\{ u^2 - 2\mathbf{u}\cdot\mathbf{v} + v^2 - \frac{(\mathbf{u}\times\mathbf{v})^2}{c^2} \right\} \tag{2.103}$$

The corresponding transformations for the components of acceleration in the primed system to those in the unprimed system are

$$a'_x = \frac{(du'_x/dt)}{(dt'/dt)} = \frac{a_x}{\gamma^3(1 - vu_x/c^2)^3} \qquad [2.104(a)]$$

$$a'_y = \frac{a_y(1 - vu_x/c^2) + vu_y a_x/c^2}{\gamma^2(1 - vu_x/c^2)^3} \qquad [2.104(b)]$$

$$a'_z = \frac{a_z(1 - vu_x/c^2) + vu_z a_x/c^2}{\gamma^2(1 - vu_x/c^2)^3} \qquad [2.104(c)]$$

The inverse transformations are obtained by changing the sign of v and interchanging the primed and unprimed quantities in the above equations. Thus,

$$a_x = \frac{a'_x}{\gamma^3(1 + u'_x/c^2)^3}, \text{ etc.} \tag{2.105}$$

From Eq. (2.103), we infer that $\mathbf{u}$ does not transform as part of a four-vector, rather it has more complex transformation properties. This is due to the reason that $d\mathbf{r}/dt$ is not a part of a four-vector, although $\mathbf{r}$ is part of a four-vector but dt is not invariant.

Example 2.8: Transformation of Lorentz contraction factor If u and u' are the velocities of a particle in the frames S and S', which are moving with velocity v relative to each other, prove that

$$\sqrt{1 - u'^2/c^2} = \frac{\sqrt{(1 - v^2/c^2)(1 - u^2/c^2)}}{(1 - vu_x/c^2)^2}$$

Solution. The velocity of a particle relative to S' is given by

$$u'^2 = u'^2_x + u'^2_y + u'^2_z$$

$$= \frac{(u_x - v)^2 + (1 - v^2/c^2)u_y^2 + (1 - v^2/c^2)u_z^2}{(1 - vu_x/c^2)^2}$$

$$= \frac{(u_x - v)^2 + (u^2 - u_x^2)\,(1 - v^2/c^2)}{(1 - vu_x/c^2)^2}$$

Therefore,

$$1 - \frac{u'^2}{c^2} = 1 - \frac{(u_x/c - v/c)^2 + (u^2/c^2 - u_x{}^2/c^2)(1 - v^2/c^2)}{(1 - v_u x/c^2)^2}$$

$$= \frac{1 - v^2/c^2 - u^2/c^2 + v^2\,u^2/c^4}{(1 - vu_x/c^2)^2}$$

$$= \frac{(1 - v^2/c^2)(1 - u^2/c^2)}{(1 - vu_x/c^2)^2}$$

Taking the square root of both the sides, we get

$$\sqrt{1 - u'^2/c^2} = \frac{\sqrt{(1 - v^2/c^2)(1 - u^2/c^2)}}{(1 - vu_x/c^2)}$$

Example 2.9 Find an expression for the absolute magnitude of relative velocity in relativistic mechanics.

Solution. The relative velocity of two particles is defined in relativistic mechanics as the velocity of one of them in the system of reference in which the other is at rest. If u_1, u_2 are the velocities of the two in the original coordinate system, then setting $u' = u_1$, $v = -u_2$ in Eqs (2.71), the components of relative velocity are given by

$$u_x = \frac{u_{1x} - u_{2x}}{1 - u_2 u_{1x}/c^2}$$

$$u_y = \frac{u_{1y}\sqrt{1 - u_x^2}}{1 - u_2 u_{1x}/c^2}$$

$$u_z = \frac{u_{1z}\sqrt{1 - u_2^2/c^2}}{1 - u_2 u_{1x}/c^2}$$

The square of absolute value of relative velocity is

$$\mathbf{u}_{rel}^2 = \frac{(\mathbf{u}_1 - \mathbf{u}_2)^2 - 1/c^2\,(\mathbf{u}_1 \times \mathbf{u}_2)^2}{(1 - \mathbf{u}_1 \cdot \mathbf{u}_2/c^2)^2}$$

As is obvious, the above expression is symmetric in u_1 and u_2, i.e., the magnitude of the relative velocity is independent of our choice of the particle with respect to which it is defined.

The equation giving the square of absolute value of relative velocity simplifies, if in the given frame, the two velocities u_1 and u_2 are either parallel or perpendicular. In the first case, we get

$$\mathbf{u}_{rel} = \frac{\mathbf{u}_1 - \mathbf{u}_2}{1 - \mathbf{u}_1 \cdot \mathbf{u}_2/c^2} \qquad \text{since } (\mathbf{u}_1 \times \mathbf{u}_2) = 0$$

and in the second case

$$\mathbf{u}_{rel} = \mathbf{u}_1\sqrt{1 - u_2^2/c^2} - \mathbf{u}_2 \qquad \text{since } (\mathbf{u}_1 \cdot \mathbf{u}_2) = 0$$

References

1. H. Fizeau, *Comp. Rend.* (1851) 33, 349.
2. F. T. Trouton and H. R. Noble, *Phil. Trans. Roy. Soc. Lond.* (1903) *A 202*, 165; *Proc. Roy. Soc. Lond.* (1903) *72*, 132.
3. A. A. Michelson and E. W. Morley, *Am. J. Sci.* (1887) *34*, 333; *Phil. Mag.* (1887) *24*, 449.
4. C. J. Cedarholm, G. F. Havens, B. L. Bland and C. H. Townes, *Phys. Rev. Lett.* (958) *1*, 342.
5. H. A. Lorentz, *Proc. Acad. Sci. Amsterdam* (1904) *6*, 809.
6. H. Poincaré, *C. R. Acad. Sci. Paris* (1905) *140*, 1504; *R. C. Circ. Mat. Palermo* (1906) *21*, 129.
7. A. Einstein, *Ann. Phys. Leipzig* (1905) *17*, 891. Also see *The Principle of Relativity*, Dover Publications Inc., New York, 1958.
8. C. Møller, *The Theory of Relativity*, Clarendon Press, Oxford, 1952.
9. A. Grünbaum, *Am. J. Phys.* (1955) *23*, 450.
10. K. Brecher, *Phys Rev. Lett.* (1977) *39*, 1051
11. T. Alväger, F. J. M. Farley, J. Kjellman and I. Walling, *Phys. Lett.* (1977) *12*, 260.
12. G. Nadeau, *Am. J. Phys.* (1962) *30*, 602. The Lorentz transformation is obtained by a vector method; for axiomatic deduction of general Lorentz transformations, consult H. M. Schwartz, *Am. J. Phys.* (1962) *30*, 697.
13. R. Weinstein, *Am. J. Phys.* (1960) *28*, 607.

14. E. M. Purcell, *Electricity and Magnetism* (1965), McGraw-Hill Publishing Company, New York.

15. A. D. Crowell, *Am. J. Phys.* (1961) *29*, 370–371, gives extension of argument to measurement of time intervals by a single observer.

16. B. Rossi and D. B. Hall, *Phys. Rev.* (1941) *59*, 223.

17. F. J. M. Farley et al., *Nouvo Cimento* (1966) *45*, 281

18. R. Mössbauer, *Z. Physik*, (1958) *151*, 124.

19. H. Hay, J. Schiffer, T. Cranshaw and P. Egelstaff, *Phys. Rev. Lett.* (1960) *4*, 165.

20. G. Saathoff et al., "Improved test of time dilation in special relativity", *Phys. Rev. Lett.* (2003) *91*, 190403.

21. Bailey et al., "Measurements of relativistic time dilation for positive and negative muons in a circular orbit", *Nature* (1977) *268*, 301.

22. Bailey et al., *Nucl. Phys. B* (1979) *150*, 1–79.

23. C. W. Sherwin, "Some recent experimental tests of the clock paradox", *Phys. Rev.* (1960) *120*, 17. Discusses the recent experimental work of Pound–Rebka and of Hay–Schiffer–Cranshaw–Egelstaff as experimental verification of the velocity dependence of the clock rates.

24. B. Liebowitz, *Am. J. Phys.* (1956) *24*, 587.

25. A. Grünbaum, *Am. J. Phys.* (1956) *24*, 588.

Problems

2.1 Find the representation of a vector in the primed frame that is $6\mathbf{i} + 7\mathbf{j}$ in the unprimed frame, the latter moving at $0.6c\,\mathbf{i}$ with respect to the primed frame. $(4.8\mathbf{i}' + 7\mathbf{j}')$

2.2 A thin rod of length $l_0 = 1$ m, as measured by a proper observer, is moving at $0.8\,c$ with respect to a second observer in a direction at 60° to its own length. What length of the rod is measured by the second observer? (91.6 cm)

2.3 Find the time difference between the transit times of two beams of light moving in opposite directions in water ($\mu = 1.33$) having velocity $v = 3$ m/s in one of the beam directions. The path length L for each beam in water as measured in the laboratory is 15 m. (10^{-15} s)

2.4 An unstable meson with mean proper lifetime of 2.2 μs has velocity 2×10^{10} cm/s.

(a) What will be its lifetime as measured by an observer in the laboratory?

(b) How far does the meson go in the laboratory before disintegrating?

2.5 Observers in inertial frames S and S' synchronize clocks at the origin. After 10 min by their respective clocks, they look through telescopes at each other. What does clock in S read for S' and vice versa, when $\gamma = 5/3$. (10.8 min)

2.6 Two electron beams travel along the same straight line but in opposite directions with velocities $v = 0.9c$ relative to the laboratory system. Find the relative velocity v of the electrons according to Newtonian mechanics. What will be the velocity measured by an observer moving with one of the electron beams? ($0.8c$;$0.994c$)

2.7 A photon is observed from a spaceship moving with a speed of $0.9c$. What is the speed of the photon with respect to the spaceship? (c)

2.8 In Fizeau's experiment, the values of the parameters are as follows: $l = 1.5$ m, $\mu = 1.33$, $\lambda = 5.3 \times 10^{-7}$ m and velocity of water = 7 m/s. A shift of 0.23 m fringe was observed from the case when water was at rest. Calculate the drag coefficient and compare it with the predicted value. (0.49;0.43)

2.9 An observer moving towards a star finds that half the radiation emitted by the star is concentrated within a cone of $\pm$ 0.1 rad. Find the speed of the observer.
(use $\cos\theta = 1 - \theta^2/2 + \cdots$ when θ is small) ($\beta = 0.995$)

2.10 According to the Lorentz–Fitzgerald contraction consequence of Lorentz transformation, length of all bodies in motion through space contracts in the direction of motion in the ratio $(1 - v^2/c^2)^{1/2}$: 1, where v is the velocity through space. A body is spherical in shape when at rest. Determine the shape of this object

(a) as measured (or observed) in the laboratory frame

(b) as photographed (or visually detected) in the laboratory frame

[(a) ellipsoidal; (b) spherical]

2.11 A UFO is moving at 7.8×10^7 m/s relative to a plane. The plane is flying at a velocity of 2.2×10^7 of light $c = 3.00 \times 10^8$ m/s. (9.43×10^7 m/s)

2.12 Two particles move towards each other with velocity v, in the reference frame of an observer. The angle between the velocity vectors is. What is the speed of one particle as viewed by the other.

$$\left[\frac{2v\sin\theta}{1 - v^2\cos^2\theta/c^2} \bigg/ \left(1 + \frac{v^2\sin^2\theta}{(c^2 - v^2\cos^2\theta)}\right)\right]$$

2.13 Twin paradox. A and B are two twins. A stays on the earth and B flies to a far-away star and back. Show that B is younger than A when they meet up again.

2.14 Consider a frame S' moving at speed v with respect to S. Consider two events at the origin of S′, separated by the time t'. Using the invariance of the square of interval, derive the time dilation result.

2.15 For an observer in a laboratory system, two photons are moving in directions opposite to each other. Show that the velocity of one photon with respect to the other is c.

2.16 Consider two particles (each with the lifetime $\tau = 2 \times 10^{-8}$ s) created simultaneously, particle 1 at point P with $\gamma = 1$ and particle 2 at P' with $\gamma = 100$ moving along PP'.

(a) Calculate what should be the distance $L = PP'$ in frame S associated with particle 1 so that both the particles decay simultaneously at P. (600 m)

(b) Calculate the distance L as calculated by observer in frame S' of particle 2. (600 m)

(c) Are the above results compatible with each other? (yes)

2.17 A person with a 'rest' heart beat of 70 beat per minute is travelling at a speed of 0.9c relative to an observer. What is the heart beat rate as measured by the observer? (30.5 per min.)

2.18 The muon is an unstable particle that decays spontaneously into an electron and two neutrinos, with the mean lifetime of 2.20 μs. If the muon moves at a speed of 0.95c, calculate the following:

(a) What is the observed lifetime of the muon? (7.046 μs)

(b) How many muons remain after travelling a distance of 3.0 km? (0.225 N_0)

Hint. If the number at $t = 0$ is N_0 the number N at time t is $N = N_0 e^{-t/\tau}$ where $\tau = 2.2$ μs is the mean lifetime of the muon.

2.19 A rod of length L_0 moves with speed v along x'-axis and makes an angle θ_0 with x'axis:

(a) Determine the length of the rod as measured by a stationary observer. $[L_0(1 + \beta^2\cos^2\theta_0)]$

(b) Determine the angle that the rod makes with the x-axis. $[\theta = \tan^{-1} \gamma \tan\theta_0]$

2.20 As observed on earth, two spaceships A and B are approaching along perpendicular directions. If A is observed by a stationary observer to have velocity $u_y = -0.90c$ and B to have velocity $u_x = +0.90c$, determine the speed of ship A as observed by the pilot of ship B. $(0.95c)$

2.21 Calculate the average speed of a μ-meson to travel a distance of 20 m before decaying. The average rest lifetime of a μ-meson is 2.2×10^{-6} s. $(1.72 \times 10^6$ m/s)

3

Mathematical Background

3.1 LINEAR ORTHOGONAL TRANSFORMATION AND ITS INVARIANTS

The central problem in the theory of relativity is to formulate the laws of physics and, in particular, the equations of the electromagnetic field, in a form which is invariant under transformations to another reference system that is moving uniformly with respect to the first. In this chapter, we will confine ourselves to systems of rectangular Cartesian coordinates.

The position of a particle P in a given Cartesian coordinate system $OXYZ$ is given by the vector

$$\mathbf{r} = x_1\mathbf{i}_1 + x_2\mathbf{i}_2 + x_3\mathbf{i}_3 \tag{3.1}$$

where $\mathbf{i}_1, \mathbf{i}_2, \mathbf{i}_3$ are respectively unit vectors along the X-, Y-, and Z- axes. These unit vectors are called *base vectors*. They define the coordinate system or basis in which x_1, x_2, x_3 are the coordinates of a particle. Moreover, the base vectors $\mathbf{i}_1, \mathbf{i}_2, \mathbf{i}_3$ are linearly independent.

Since

$$\mathbf{i}_j \cdot \mathbf{i}_k = \delta_{jk}$$

the coordinates of P in the system S are

$$x_k = \mathbf{r} \cdot \mathbf{i}_k \tag{3.2}$$

Supposing $\mathbf{i}'_1, \mathbf{i}'_2, \mathbf{i}'_3$ are the base vectors of a second rectangular system S', whose origin O' coincides with O, which thus differs from S only by a rotation.

Since

$$\mathbf{r} = x'_1\mathbf{i}'_1 + x'_2\mathbf{i}'_2 + x'_3\mathbf{i}'_3$$

so the coordinates of P with respect to S' are

$$\begin{aligned} x'_j &= \mathbf{r} \cdot \mathbf{i}'_j \\ &= x_1\mathbf{i}_1 \cdot \mathbf{i}'_j + x_2\mathbf{i}_2 \cdot \mathbf{i}'_j + x_3\mathbf{i}_3 \cdot \mathbf{i}'_j \end{aligned} \tag{3.3}$$

Obviously the coefficients

$$a_{jk} = \mathbf{i}'_j \cdot \mathbf{i}_k \tag{3.4}$$

of the linear form are the direction cosines of the coordinate axes of S' with respect to the axes of S. When the rectangular coordinate system is rotated, the coordinates of a point in the new system related to those in the old one by the linear transformation are

$$x'_j = \sum_{k=1}^{3} a_{jk} x_k \qquad (j = 1, 2, 3) \tag{3.5}$$

But, the magnitude of the vector **r** is independent of the orientation of the coordinate system, hence

$$\sum_{j=1}^{3}(x_j)^2=\sum_{j=1}^{3}(x'_j)^2$$

Now,

$$\sum_{j=1}^{3}(x'_j)^2=\sum_{j=1}^{3}\left(\sum_{i=1}^{3}a_{ji}x_i\right)\left(\sum_{k=1}^{3}a_{jk}x_k\right)$$

$$=\sum_{j=1}^{3}\sum_{k=1}^{3}x_i x_k\left(\sum_{k=1}^{3}a_{ji}a_{jk}\right)$$

whence it is clear that

$$\sum_{i=1}^{3}a_{ji}a_{jk}=\delta_{ik}=\begin{cases}1 \text{ if } i=k\\0 \text{ if } i\neq k\end{cases}\tag{3.6}$$

Equation (3.6) expresses the relations between the cosines of the angles between the coordinate axes in order that these be rectangular and are hence known as conditions of orthogonality. The corresponding transformation (3.5) when subjected to the condition of orthogonality (3.6) is called the orthogonal transformation.

It is shown in the next section that the square of the determinant $|a_{jk}|$ is equal to unity,

$$|a_{jk}|=\pm 1$$

$|a_{jk}|=+1$, that is, the orthogonal transformation whose determinant is equal to +1 represents a rotation, whereas $|a_{jk}|=-1$, corresponds to an inversion followed by a rotation.

Since the determinant of an orthogonal transformation does not vanish, the inverse transformation of expressing x_k in terms of x'_j is also definable. Thus,

$$x_k=\mathbf{r}\cdot\mathbf{i}_k$$

$$=x'_1\mathbf{i}'_1\cdot\mathbf{i}_k+x'_2\mathbf{i}'_2\cdot\mathbf{i}_k+x'_3\mathbf{i}'_3\cdot\mathbf{i}_k$$

or

$$x_k=\sum_{j=1}^{3}a_{jk}x'_j\tag{3.7}$$

As this is also an orthogonal transformation, the corresponding condition of orthogonality is easily derived as

$$\sum_{i=1}^{3}a_{ij}a_{kj}=\delta_{ik}\tag{3.8}$$

Let us examine the effect of the process of rotation on a vector and a scalar. If **A** is any fixed vector in space, then

$$\mathbf{A}=\sum_{k=1}^{3}A_k\mathbf{i}_k=\sum_{k=1}^{3}A'_k\mathbf{i}'_k\tag{3.9}$$

The components A'_j of this vector with respect to the coordinate system S' are given by

$$
\begin{aligned}
A'_j &= \mathbf{A}\cdot\mathbf{i}'_j \\
&= \sum_{k=1}^{3} A_k \mathbf{i}_k\cdot\mathbf{i}'_j = \sum_{k=1}^{3} a_{jk} A_k
\end{aligned}
$$

that shows that A'_j upon rotation transform like the coordinates of a point. Thus, in order to decide that the three scalars A_1, A_2, A_3 constitute the three components of a vector, these must transform like the coordinate of a point.

Scalars are of two types, variant and invariants. Those quantities that are independent of the orientation of the coordinate system, such as temperature and work are called invariant scalars whereas quantities which have only magnitude but transform with the coordinate system are termed variant scalars. The coordinates of a point and components of a vector are examples of variant scalars.

If **A** and **B** are two vectors, their scalar product **A·B** is a scalar. In order to decide whether it belongs to invariant type, we will examine if it undergoes orthogonal transformation. Thus,

$$
\begin{aligned}
\mathbf{A}\cdot\mathbf{B} &= \sum_{k=1}^{3} A_k B_k \\
&= \sum_{k=1}^{3}\left(\sum_{j=1}^{3} a_{jk} A'_j\right)\left(\sum_{i=1}^{3} a_{ik} B'_i\right) \\
&= \sum_{j=1}^{3} A'_j \cdot B'_j \qquad (3.10)
\end{aligned}
$$

since $\sum_{k=1}^{3} a_{jk} a_{ik} = \delta_{ij}$.

Thus the scalar product of two vectors is an invariant scalar.

Let us further consider the nature of gradient of an invariant scalar. Defining

$$B_i = \frac{\partial \phi}{\partial x_i} \qquad (i = 1, 2, 3) \qquad (3.11)$$

so,

$$B'_i = \frac{\partial \phi}{\partial x'_i} = \sum_{k=1}^{3} \frac{\partial \phi}{\partial x_k}\frac{\partial x_k}{\partial x'_i} = \sum_{k=1}^{3} a_{ik} B_k$$

where $a_{ik} = \dfrac{\partial x_k}{\partial x'_i}$.

with the consequence that B_i transforms like the components of a vector and thus we infer that the gradient of ϕ

$$\nabla\phi = \sum_{k=1}^{3} \frac{\partial \phi}{\partial x_k}\mathbf{i}_k \qquad (3.12)$$

at a point is a fixed vector associated with that point.

Next, we examine the case of divergence of a vector.

Defining $$B_i = \frac{\partial A_i}{\partial x_i} \qquad (i = 1, 2, 3)$$

where A_i are the rectangular components of a vector **A**.

Now, $$B'_i = \frac{\partial A'_i}{\partial x'_i}$$

$$= \sum_{k=1}^{3} \frac{\partial A'_i}{\partial x_k} \frac{\partial x_k}{\partial x'_i}$$

$$= \sum_{k=1}^{3} a_{ik} \frac{\partial A'_i}{\partial x_k}$$

$$= \sum_{k=1}^{3} \sum_{j=1}^{3} a_{ik} a_{ij} \frac{\partial A_j}{\partial x_k}$$

But, $$\sum_{i=1}^{3} a_{ik} a_{ij} = \delta_{kj}$$

so, $$\sum_{i=1}^{3} B'_i = \sum_{k=1}^{3} \sum_{j=1}^{3} \left(\sum_{i=1}^{3} a_{ik} a_{ij} \right) \frac{\partial A_j}{\partial x_k}$$

$$= \sum_{j=1}^{3} B_j \qquad (3.13)$$

The divergence of a vector is invariant to an orthogonal transformation of the coordinate system. Now that the gradient of an invariant scalar is a vector and the divergence of a vector is invariant, it follows that

$$\nabla \cdot \nabla \phi = \nabla^2 \phi$$

$$= \sum_{i=1}^{3} \frac{\partial^2 \phi}{\partial x_i^2} \qquad (3.14)$$

called Laplacian, is invariant under orthogonal transformation.

Example 3.1 Show that a four-dimensional divergence and d'Alembert's operator are invariant under Lorentz transformation.

Solution. According to Lorentz transformation

$$x'_\mu = \sum_\nu \alpha_{\mu\nu} x_\nu$$

$$\frac{\partial x'_\mu}{\partial x_\nu} = \alpha_{\mu\nu}$$

and for the components of the vector **b**

$$b'_\mu = \sum_\nu \alpha_{\nu\mu} b_\nu;\; b_\mu = \sum_\nu \alpha_{\nu\mu} b'_\nu$$

Firstly, we show that a four-gradient is transformed as a vector. Consider the function $\phi = \phi(x'_0, x'_1, x'_2, x'_3)$ or briefly $\phi = \phi(x'_\mu)$. Then according to the differentiation rule for a composite function, we have

$$\frac{\partial \phi}{\partial x_\nu} = \sum_\mu \frac{\partial \phi}{\partial x'_\mu}\frac{\partial x'_\mu}{\partial x_\nu} = \sum_\mu \alpha_{\mu\nu}\frac{\partial \phi}{\partial x'_\mu}$$

Furthermore, the four-divergence of a four-vector is

$$\Box \cdot \mathbf{A} = \sum_\nu \frac{\partial A_\nu}{\partial x_\nu} = \sum_{\mu,\nu} \frac{\partial A_\nu}{\partial x'_\mu}\frac{\partial x'_\mu}{\partial x_\nu}$$

$$= \sum_{\mu,\nu,\sigma} \alpha_{\mu\nu}\,\alpha_{\nu\sigma}\frac{\partial A'_\sigma}{\partial x'_\mu}$$

$$= \sum_{\mu,\sigma} \delta_{\mu\sigma}\frac{\partial A'_\sigma}{\partial x'_\mu} = \sum_\mu \frac{\partial A'_\mu}{\partial x'_\mu}$$

D'Alembert's operator when applied to the function ϕ

$$\Box^2\phi = \nabla^2\phi - \frac{1}{c^2}\frac{\partial^2 \phi}{\partial t^2} = \frac{\partial^2 \phi}{\partial x^2} + \frac{\partial^2 \phi}{\partial y^2} + \frac{\partial^2 \phi}{\partial z^2} - \frac{1}{c^2}\frac{\partial^2 \phi}{\partial t^2}$$

can be written as

$$\sum_\mu \frac{\partial^2 \phi}{\partial x_\mu^2} = \frac{\partial^2 \phi}{\partial x_1^2} + \frac{\partial^2 \phi}{\partial x_2^2} + \frac{\partial^2 \phi}{\partial x_3^2} + \frac{\partial^2 \phi}{\partial (ict)^2}$$

However, the expression $\sum_\mu \frac{\partial^2 \phi}{\partial x_\mu^2}$ is the divergence of a gradient. Let $A_\mu = \frac{\partial \phi}{\partial x_\mu}$, then

$$\operatorname{div}\mathbf{A} = \sum_\mu \frac{\partial}{\partial x_\mu}\left(\frac{\partial \phi}{\partial x_\mu}\right) = \sum_\mu \frac{\partial^2 \phi}{\partial x_\mu^2}$$

As shown earlier, the divergence of a four-vector is an invariant of the Lorentz transformation, therefore

$$\sum_\mu \frac{\partial^2 \phi}{\partial x_\mu^2} = \sum_\mu \frac{\partial^2 \phi}{\partial x'^2_\mu}$$

3.2 MINKOWSKI'S FOUR-DIMENSIONAL SPACE–TIME WORLD

The negative results of the Michelson–Morley experiment point out that velocity does not behave exactly like the mathematical object called vector. The resultant of two parallel vectors of magnitude c and v is, by the vector operation of addition, equal to a vector of magnitude $v + c$. The resultant vector, which results from the addition of the velocity of light to the velocity of moving reference system, still gives the same velocity of light, and so it does not behave like a mathematical vector.

Minkowski[1] imparted great elegance to the theory by introducing a fictitious construct, four-dimensional space called *Minkowski world*. The basic elements of the physical world are *events*, which are described by the time when occurred and by the space where occurred. Thus an event has four coordinates for its location, i.e., t, x, y, z. The main crux of Minkowski's theory is that these four numbers must be treated as an ordered set of four components of a vector and a suitable algebra of such four-dimensional vectors must be built to describe the relationships between them.

Consider two reference systems S and S', the x-axis parallel to the x'-axis and the origins O and O' coinciding at $t = 0$. The system S' moves with velocity v with respect to S. Supposing that at time $t = 0$, a spherical electromagnetic wave is sent into free space from the common origin. At any subsequent time, the equations of wave front given in the two systems S and S' are

$$c^2t^2 = x^2 + y^2 + z^2 \tag{3.15}$$

$$c^2t'^2 = x'^2 + y'^2 + z'^2 \tag{3.16}$$

These are the equations of a sphere moving out with speed c in all directions. (ct, x, y, z) are the time and coordinates measured by the observer in S whereas (ct', x', y', z'), the time and coordinates measured by the observer in S'. These two conflicting descriptions of the same physical phenomenon can be reconciled if we adopt the following linear transformations

$$t' = \gamma\left(t - \frac{vx}{c^2}\right) \qquad [3.17(a)]$$

$$x' = \gamma(x - vt) \qquad [3.17(b)]$$

$$y' = y; z' = z \qquad [3.17(c)]$$

where $\gamma = \left(1 - \frac{v^2}{c^2}\right)^{-\frac{1}{2}}$.

The mathematical implications of these transformations suggest that time cannot be considered a scalar under it and the three space coordinates do not behave like the components of a classical three-dimensional vector. But when $v << c$, Lorentz contraction factor $\gamma \to 1$ and the usual classical meanings are ascribed to t, x, y, and z.

We see that the expression $c^2t^2 - x^2 - y^2 - z^2$ is invariant under Lorentz transformation. This is called the square length of the interval, whose four components are (ct, x, y, z). Representing the coordinates of a point P by x_0, x_1, x_2, x_3 in a four-dimensional continuum, the location of a point P, called *world point*, with respect to a fixed origin O is determined by

$$R = \sum_{\mu=0}^{3} x_\mu i_\mu \qquad i_\mu . i_\nu = \delta_{\mu\nu} \tag{3.18}$$

The linear transformation

$$x'_\mu = \sum_{\mu=0}^{3} \alpha_{\mu\nu} x_\nu \qquad [3.19(a)]$$

will be orthogonal if the coefficients are subjected to the following condition

$$\sum_{\mu=0}^{3} \alpha_{\mu\rho}\alpha_{\mu\nu} = \delta_{\rho\nu} \qquad [3.19(b)]$$

3.2.1 Notation

We shall denote the components of three-dimensional vectors by Latin indices *i*, *j*, *k* whereas the Greek indices μ, ν, ρ, λ will be used to denote the components of a four-vector. In accordance with the usual convention of Tensor analysis, the repetition of an index implies summation over the range of the index. Thus, in a three-dimensional space

$$\mathbf{A}\cdot\mathbf{B} = A_i B_i = A_1 B_1 + A_2 B_2 + A_3 B_3 \qquad [3.20(a)]$$

whereas when *A* and *B* stand for four-vectors, we have

$$AB = A_\mu B_\mu = A_0 B_0 - A_1 B_1 - A_2 B_2 - A_3 B_3 \qquad [3.20(b)]$$

3.3 LORENTZ TRANSFORMATION AS ROTATION IN FOUR-SPACE

We will again show that classical transformation equations (Galilean) can be replaced by Lorentz transformations that are based on the invariance of the velocity of light and not on the assumption of universal time and absolute space (invariant length of scale). In deriving these equations, we assume the principle of relativity as fundamental, implying that all inertial frames are equivalent. Furthermore, it is assumed that space is homogeneous, i.e., all points of space are equivalent. The resulting equations must be linear equations.

The postulate of the constancy of velocity of light *c* will be obeyed by the group of transformations that leave the equation of wave front invariant

$$x^2 + y^2 + z^2 - c^2t^2 = x'^2 + y'^2 + z'^2 - c^2t'^2 \qquad (3.21)$$

The space in which the line element, Eq. (3.21), is written is pseudo-Euclidean. Euclidean manifold is a manifold with all coefficients equal to 1, i.e., +1, +1, +1, +1 in four dimensions. These are called the signature of space. This may be achieved by choosing

$$x_0 = ict$$

$$x_1 = x$$

$$x_2 = y$$

$$x_3 = z$$

Therefore,

$$x_0^2 + x_1^2 + x_2^2 + x_3^2 = x_0'^2 + x_1'^2 + x_2'^2 + x_3'^2 \qquad (3.22)$$

But the above equation is invariant to the group of rotations in four-space and hence we construe that transformation that connects the coordinates of an event in *S* system to the coordinates of the same event in *S′*, corresponding to a rotation in a four-dimensional space x_0, x_1, x_2, x_3. The four-space consists of three dimensions of ordinary space with a fourth imaginary dimension proportional to time. This space is called *world space or Minkowski space*.

Writing the transformation in the component form, we get

$$x_0' = \alpha_{00}x_0 + \alpha_{01}x_1 + \alpha_{02}x_2 + \alpha_{03}x_3 \qquad [3.23(a)]$$

$$x_1' = \alpha_{10}x_0 + \alpha_{11}x_1 + \alpha_{12}x_2 + \alpha_{13}x_3 \qquad [3.23(b)]$$

$$x_2' = \alpha_{20}x_0 + \alpha_{21}x_1 + \alpha_{22}x_2 + \alpha_{23}x_3 \qquad [3.23(c)]$$

$$x_3' = \alpha_{30}x_0 + \alpha_{31}x_1 + \alpha_{32}x_2 + \alpha_{33}x_3 \qquad [3.23(d)]$$

or

$$x_\mu' = \sum_{\nu=0}^{3} \alpha_{\mu\nu} x_\nu \qquad (3.24)$$

where the matrix elements satisfy the orthogonality condition

$$\sum_{\mu=0}^{3} \alpha_{\mu\rho}\alpha_{\mu\nu} = \delta_{\rho\nu} \qquad (\rho, \nu = 0, 1, 2, 3) \qquad [3.24(a)]$$

with the determinant $|\alpha_{\mu\nu}| = +1$.

Now

$$x' = \mathcal{L}(\alpha)x \qquad (3.25)$$

where $\mathcal{L}(\alpha)$, called the Lorentz transformation matrix, is

$$\mathcal{L}(\alpha) = \begin{pmatrix} \alpha_{00} & \alpha_{01} & \alpha_{02} & \alpha_{03} \\ \alpha_{10} & \alpha_{11} & \alpha_{12} & \alpha_{13} \\ \alpha_{20} & \alpha_{21} & \alpha_{22} & \alpha_{23} \\ \alpha_{30} & \alpha_{31} & \alpha_{32} & \alpha_{33} \end{pmatrix}$$

It may be remarked that unlike the elements of the spatial orthogonal transformation, the matrix elements $\alpha_{\mu\nu}$ are not all real, since all the components (x_0, x_1, x_2, x_3), of the four-vector x_μ are not all real. The elements α_{i0} $(i = 1, 2, 3)$ must be imaginary, so that the coordinates x_1', x_2', x_3' remain real. In addition, x_0' being imaginary, it is essential that α_{0i} $(i = 1, 2, 3)$ be all imaginary, whereas α_{00} must be real.

As the distances perpendicular to the direction of motion are not affected, the rotation leaves the coordinates x_2, x_3 unchanged, thus giving

$$x_2' = x_2$$

$$x_3' = x_3 \qquad (3.26)$$

In view of Eq. (3.26), the transformation matrix reduces to

$$\mathcal{L}(\alpha) = \begin{pmatrix} \alpha_{00} & \alpha_{01} & 0 & 0 \\ \alpha_{10} & \alpha_{11} & 0 & 0 \\ 0 & 0 & 1 & 0 \\ 0 & 0 & 0 & 1 \end{pmatrix} \qquad (3.27)$$

which written out in the component form gives

$$x'_0 = \alpha_{00}x_0 + \alpha_{01}x_1 \qquad [3.28(a)]$$

$$x'_1 = \alpha_{10}x_0 + \alpha_{11}x_1 \qquad [3.28(b)]$$

$$x'_2 = x_2 \qquad [3.28(c)]$$

$$x'_3 = x_3 \qquad [3.28(d)]$$

The orthogonality condition thus gives

$$x'^2_0 + x'^2_1 = x^2_0 + x^2_1 \qquad (3.29)$$

which obviously holds if the transformation is in $x_0 - x_1$ plane. Representing x'_0, x'_1 by x' and x_0, x_1 by x, then

$$x' = \alpha x \qquad (3.30)$$

where

$$\alpha = \begin{pmatrix} \alpha_{00} & \alpha_{01} \\ \alpha_{10} & \alpha_{11} \end{pmatrix} \qquad (3.31)$$

Transposing Eq. (3.30), we get

$$x'^T = x^T \alpha^T \qquad (3.32)$$

Therefore the condition of orthogonality gives

$$x'^T x' = x^T \alpha^T \alpha x = x^T x \qquad (3.33)$$

whence I is a 2 × 2 unit matrix

$$\begin{pmatrix} \alpha_{00} & \alpha_{10} \\ \alpha_{01} & \alpha_{11} \end{pmatrix} \begin{pmatrix} \alpha_{00} & \alpha_{01} \\ \alpha_{10} & \alpha_{11} \end{pmatrix} = I = \begin{pmatrix} 1 & 0 \\ 0 & 1 \end{pmatrix}$$

Thus we obtain

$$\alpha^2_{00} + \alpha^2_{01} = 1$$

$$\alpha^2_{10} + \alpha^2_{11} = 1 \qquad (3.34)$$

$$\alpha_{00}\alpha_{10} + \alpha_{01}\alpha_{11} = 0$$

There are four unknowns to be determined from three equations, so we require a fourth equation to determine these uniquely. An additional condition is supplied by the observation that the origin of the primed system S' is moving uniformly along the x_1-axis, so that at time t, its x_1 coordinate is vt. Thus,

$$x_1 = vt = -i\frac{v}{c}x_0 = -i\beta x_0 \qquad (3.35)$$

where $\beta = \frac{v}{c}$. Further, the origin of the primed system S' referred to itself is

$$x_1' = \alpha_{10}x_0 + \alpha_{11}x_1 = 0$$

or

$$x_1' = (\alpha_{10} - i\beta\alpha_{11})x_0 = 0 \tag{3.36}$$

which gives

$$\alpha_{10} = i\beta\alpha_{11}$$

Substituting Eq. (3.36) into the second equation of Eq. (3.34), we get

$$\alpha_{11}^2(1 - \beta^2) = 1$$

or

$$\alpha_{11} = \frac{1}{\sqrt{1 - \beta^2}} \tag{3.38}$$

Thus

$$\alpha_{10} = \frac{i\beta}{\sqrt{1 - \beta^2}} \tag{3.39}$$

Lastly, let us derive α_{01} from the third equation of Eq. (3.34)

$$\alpha_{01} = -\frac{\alpha_{10}}{\alpha_{11}}\alpha_{00}$$

$$= -i\beta\alpha_{00}$$

which when substituted in the first equation of Eq. (3.34) yields

$$\alpha_{00}^2(1 - \beta^2) = 1$$

or

$$\alpha_{00} = \frac{1}{\sqrt{1 - \beta^2}} \tag{3.40}$$

Therefore,

$$\alpha_{01} = -\frac{i\beta}{\sqrt{1 - \beta^2}} \tag{3.41}$$

The Lorentz transformation matrix $\mathcal{L}$ finally becomes

$$\mathcal{L} = \begin{pmatrix} \frac{1}{\sqrt{1-\beta^2}} & \frac{-i\beta}{\sqrt{1-\beta^2}} & 0 & 0 \\ \frac{i\beta}{\sqrt{1-\beta^2}} & \frac{1}{\sqrt{1-\beta^2}} & 0 & 0 \\ 0 & 0 & 1 & 0 \\ 0 & 0 & 0 & 1 \end{pmatrix} \tag{3.42}$$

Writing the Lorentz transformation in the matrix form, we get

$$\begin{pmatrix} x_0' \\ x_1' \\ x_2' \\ x_3' \end{pmatrix} = \begin{pmatrix} \frac{1}{\sqrt{1-\beta^2}} & \frac{-i\beta}{\sqrt{1-\beta^2}} & 0 & 0 \\ \frac{i\beta}{\sqrt{1-\beta^2}} & \frac{1}{\sqrt{1-\beta^2}} & 0 & 0 \\ 0 & 0 & 1 & 0 \\ 0 & 0 & 0 & 1 \end{pmatrix} \begin{pmatrix} x_0 \\ x_1 \\ x_2 \\ x_3 \end{pmatrix} \tag{3.43}$$

and writing it in the component form, we obtain

$$x_0' = \frac{x_0 - i\beta x_1}{\sqrt{1-\beta^2}} \qquad [3.44(a)]$$

$$x_1' = \frac{x_1 + i\beta x_0}{\sqrt{1-\beta^2}} \qquad [3.44(b)]$$

$$x_2' = x_2 \qquad [3.44(c)]$$

$$x_3' = x_3 \qquad [3.44(d)]$$

or

$$t' = \frac{t - \frac{v}{c^2}x}{\sqrt{1-\beta^2}} \qquad [3.45(a)]$$

$$x' = \frac{x - vt}{\sqrt{1-\beta^2}} \qquad [3.45(b)]$$

$$y' = y \qquad [3.45(c)]$$

$$z' = z \qquad [3.45(d)]$$

The inverse transformations are obtained by replacing v by $-v$ and changing primed quantities into unprimed and vice versa. Thus,

$$t = \frac{t' + \frac{v}{c^2}x}{\sqrt{1-\beta^2}} \qquad [3.46(a)]$$

$$x = \frac{x' + vt}{\sqrt{1-\beta^2}} \qquad [3.46(b)]$$

$$y = y' \qquad [3.46(c)]$$

$$z = z' \qquad [3.46(d)]$$

Example 3.2 Show that for any Lorentz transformation, the determinant can be only ±1.

Solution. We have the linear transformation

$$x_\mu' = \sum_{\nu=0}^{3} \alpha_{\mu\nu} x_\nu$$

which will be orthogonal if the coefficients are subjected to the following condition

$$\sum_\mu \alpha_{\mu\rho}\alpha_{\mu\nu} = \delta_{\rho\nu}$$

Since Lorentz transformation is an orthogonal transformation in Minkowski space, we have

$$dx_\mu' = \sum_\nu \alpha_{\mu\nu} dx_\nu = \sum_\lambda \alpha_{\mu\lambda} dx_\lambda$$

and

$$\sum_\mu dx_\mu'^2 = \sum_\nu dx_\nu^2$$

or
$$\sum_{\mu}\left[\left(\sum_{\nu}\alpha_{\mu\nu}dx_{\nu}\right)\left(\sum_{\lambda}\alpha_{\mu\lambda}dx_{\lambda}\right)\right]=\sum_{\nu}dx_{\nu}^{2}$$

or
$$\sum_{\mu}\alpha_{\mu\lambda}\alpha_{\mu\nu}=\delta_{\nu\lambda}$$

Denoting the transposed matrix of α by α^T, so that

$$\alpha_{\mu\nu}=\alpha^{T}_{\nu\mu}$$

Then
$$\sum_{\mu}\alpha^{T}_{\nu\mu}\alpha_{\mu\lambda}=\delta_{\nu\lambda}$$

which can be written as

$$\alpha^{T}\alpha=1$$

Since the determinant of matrix is not affected by the interchange of columns and rows, we have

$$\det\alpha^{T}=\det\alpha$$

or
$$\det\alpha=\pm1$$

3.4 FOUR-VECTORS

In four-vector calculus, equations are form invariant under general Lorentz transformations, i.e., whose equations have precisely the property required by the relativity principle of all physical laws. The form of the new or proposed law assists us immediately to recognize whether a given law is Lorentz invariant. However, some Lorentz-invariant laws require four-tensors for their expression.

A four-vector is defined as a set of four variant scalars (x^0, x^1, x^2, x^3), which transform like (dx^0, dx^1, dx^2, dx^3) or which is the same thing as the components of a point under rotation. Thus if the Lorentz transformation is given by the equation

$$x'^{\mu}=\sum_{\nu=0}^{3}\alpha^{\mu}_{\nu}x^{\nu} \qquad (\mu=0,1,2,3) \tag{3.47}$$

then the four-vectors A^{μ} are A'^{μ} related by

$$A'^{\mu}=\sum_{\nu=0}^{3}\alpha^{\mu}_{\nu}A^{\nu} \qquad (\mu=0,1,2,3) \qquad [3.48(a)]$$

and the inverse transformation is

$$A^{\mu}=\sum_{\nu=0}^{3}\overline{\alpha^{\mu}_{\nu}}A'^{\nu} \qquad (\mu=0,1,2,3) \qquad [3.48(b)]$$

where $\overline{\alpha^{\mu}_{\nu}}$ are the coefficients of inverse transformation.

The sum of two four-vectors is defined as

$$A+B=(A^0+B^0, A^1+B^1, A^2+B^2, A^3+B^3) \tag{3.49}$$

in analogy with the addition of two three-dimensional vectors.

The scalar product of two four-vectors is

$$AB = A^0B^0 - A^1B^1 - A^2B^2 - A^3B^3 \tag{3.50}$$

From Eq. (3.50), we see that the scalar product of two four-vectors is not an obvious generalization of the definition of scalar product of two three-vectors; since the space-like components have a negative sign. Its mathematical implication is that the length of a vector can be zero, when neither of the components is zero. Such a four-vector is called a *null* vector. Thus from Eq. (3.50), when $A = B$ we have

$$A = \sqrt{A.A} = 0$$

or

$$(A^0)^2 = (A^1)^2 + (A^2)^2 + (A^3)^2 \tag{3.51}$$

The displacement between two events on the path of a photon is a null vector.

If $(A^0)^2 > (A^1)^2 + (A^2)^2 + (A^3)^2$ the four-vector A has a real length and A is called *time-like;* whereas if $(A^0)^2 < (A^1)^2 + (A^2)^2 + (A^3)^2$, then A is imaginary and the four-vector A is called *space-like.*

The scalar product of two four-vectors is invariant and the condition on the coefficients of the linear transformation to be observed is

$$\sum_{\mu,\nu=0}^{3} g_{\mu\nu} A^\mu B^\nu = \sum_{\mu,\nu=0}^{3} g_{\mu\nu} A'^\mu B'^\nu \tag{3.52}$$

Substituting the values for A'^μ and B'^μ (from Eq. [3.48(a)]) in the above equation, we get

$$\sum_{\mu,\nu=0}^{3} g_{\mu\nu} A^\mu B^\nu = \sum_{\mu,\nu=0}^{3} g_{\mu\nu} \sum_{\rho=0}^{3} \alpha^\mu_\rho A^\rho \sum_{\sigma=0}^{3} \alpha^\nu_\sigma B^\sigma \tag{3.53}$$

Exchanging the places of the dummy indices μ and ρ, ν and σ on the right-hand side, we get

$$\sum_{\mu,\nu=0}^{3} g_{\mu\nu} A^\mu B^\nu = \sum_{\mu,\nu=0}^{3} A^\mu B^\nu \sum_{\rho,\sigma=0}^{3} g_{\rho\sigma} \alpha^\rho_\mu \alpha^\sigma_\nu \tag{3.54}$$

which leads to the condition on the coefficients

$$\sum_{\rho,\sigma=0}^{3} g_{\rho\sigma} \alpha^\rho_\mu \alpha^\sigma_\nu = g_{\mu\nu} \tag{3.55}$$

The square of the length of a four-vector is given by $A^\mu A_\mu$

$$A.A = \sum_\mu A^\mu A_\mu = (A^0)^2 - \mathbf{A}^2 \tag{3.56}$$

The following are a few examples of four-vectors:

1. The four-vector $x^\mu(x^0, x^1, x^2, x^3)$, which gives rise to the invariant interval ds, is defined as

$$ds^2 = c^2 (dt)^2 - [(dx^1)^2 + (dx^2)^2 + (dx^3)^2] \tag{3.57}$$

2. The four-velocity $U^\mu(\gamma c, \gamma \mathbf{v})$, where $\mathbf{v}$ is the classical velocity and γ, the Lorentz contraction factor. The invariant quantity is

$$\begin{aligned}\sum_\mu U^\mu U_\mu &= (U^0)^2 - U^2 \\ &= \gamma^2 c^2 - \gamma^2 v^2 = c^2\end{aligned} \tag{3.58}$$

3. The four-momentum $p^{\mu} = \left(\frac{\mathcal{E}}{c}, \mathbf{p}\right)$. The associated energy–momentum invariant is given by

$$p^{\mu}p_{\mu} = \mathcal{E}^2/c^2 - \mathbf{p}^2 \tag{3.59}$$

where $p = |\mathbf{p}|$.

4. The four-vector current $J^{\mu} = (c\rho, \mathbf{J})$ (3.60)
ρ is the electric charge density and **J**, the electric current density.

5. The four-vector potential $A^{\mu} = (\phi, \mathbf{A})$ (3.61)
where ϕ and **A** the scalar and vector potentials of electrodynamics, respectively.

6. The four-dimensional gradient $\nabla^* = \left(\frac{1}{c}\frac{\partial}{\partial t}, -\nabla\right)$ (3.62)
and its square is

$$\begin{aligned}\nabla^{*2} &= \sum_{\mu} \frac{\partial^2}{\partial x^{\mu}\,\partial x_{\mu}} \\ &= \frac{1}{c^2}\frac{\partial^2}{\partial t^2} - \frac{\partial^2}{\partial x^j\,\partial x_j} \\ &= \frac{1}{c^2}\frac{\partial^2}{\partial t^2}, -\nabla^2\end{aligned} \tag{3.63}$$

7. The four-wave vector

$$k^{\mu} = \left(\frac{\omega}{c}, \mathbf{k}\right) \tag{3.64(a)}$$

where ω is the angular frequency and **k**, the wave vector of a wave. It is a null vector because

$$k_{\mu}k^{\mu} = \frac{\omega^2}{c^2} - k_x^2 - k_y^2 - k_z^2 = 0 \tag{3.64(b)}$$

We assumed tacitly that $\frac{\partial}{\partial x^{\mu}}$ and j^{μ} are four-vectors. We set below to prove these statements. Let us take the case of four-dimensional gradient first.

We express $\frac{\partial}{\partial x'^{\mu}}$ through the chain rule of differentiation as

$$\frac{\partial}{\partial x'^{\mu}} = \sum_{\nu} \frac{\partial x^{\nu}}{\partial x'^{\mu}} \frac{\partial}{\partial x^{\nu}} \tag{3.65}$$

The inverse transformation from Eq. [3.48(b)] may be written as

$$x^{\nu} = \sum_{\mu} \overline{\alpha}^{\nu}_{\mu} x'^{\mu} \tag{3.66}$$

which gives

$$\frac{\partial x^{\nu}}{\partial x'^{\mu}} = \overline{\alpha}^{\nu}_{\mu} \tag{3.67}$$

Equation (3.65) becomes

$$\frac{\partial}{\partial x'^{\mu}} = \sum_{\nu} \overline{\alpha}^{\nu}_{\mu} \frac{\partial}{\partial x^{\nu}} \tag{3.68}$$

On comparing Eq. (3.68) with Eq. [3.48(a)], we infer that ∇^* transforms like a four-vector.

Next, we show that j^μ is a four-vector. We assume that a charge element dq is at rest in a frame S, and if it is contained in a volume element dV, then

$$dq = \rho_0 dV \tag{3.69}$$

ρ_0 is the volume density of charge. Each fundamental particle like an electron, proton, etc. has an elementary charge e associated with it. In addition, the charge e has discreet magnitude and the total charge can be determined by counting the number of particles. As the total number of particles cannot depend on the state of motion of the observer, it is concluded that electric charge is Lorentz invariant. As charge is conserved, the same charge when viewed from another frame S' is

$$dq = \rho_0 dV = \rho dV' = dq' \tag{3.70}$$

where ρ is the charge density in S'.

The volume elements $dV = dx^1\, dx^2\, dx^3$ and $dV' = dx'^1\, dx'^2\, dx'^3$ are related to each other as

$$dV' = \sqrt{1-\beta^2}\, dV \tag{3.71}$$

so that

$$\begin{aligned}\rho_0 dV &= \rho dV' \\ &= \rho\, dx^1\, dx^2\, dx^3 \sqrt{1-\beta^2} \\ &= \rho dV \sqrt{1-\beta^2}\end{aligned} \tag{3.72}$$

which gives

$$\rho = \frac{\rho_0}{\sqrt{1-\beta^2}} \tag{3.73}$$

Thus, ρ transforms like dx^0, i.e., like the time component of a four-vector.

The components of the current density $\mathbf{J} = \rho \mathbf{u}$ are

$$j_k = \rho u_k = \rho \frac{dx^k}{dt} \tag{3.74}$$

It is obvious that j_k behaves like dx^k ($k = 1, 2, 3$), i.e., like the space components of a four-vector. Multiplying ρ by c, so that it has the same dimensions as j_k, we conclude that $j^\mu = (c\rho, \mathbf{J})$ is a four-vector.

3.4.1 Transformation of a Four-Vector

In Newtonian mechanics, all vectors transform under Galilean transformation like the position vector and undergo the same transformation. However, in special theory of relativity, all four-vectors transform under Lorentz transformation like the time–space four-vector.

Let $A^\mu(A^0, A^1, A^2, A^3)$ be a four-vector in frame S, so that A^0 is a Galilean scalar and A^1, A^2, A^3 are the components of Galilean vector $\mathbf{A}$. Let $A'^\mu(A'^0, A'^1, A'^2, A'^3)$ denote the same four-vector in S'. If

$\mathbf{A}_{\parallel}$ and $\mathbf{A}_{\perp}$ are the components of **A** parallel and perpendicular, respectively, to the relative velocity $\boldsymbol{\beta}$, then we have

$$A'^0 = \gamma(A^0 - \boldsymbol{\beta}\cdot\mathbf{A}) \tag{3.75}$$

$$\mathbf{A}'_{\parallel} = \gamma(\mathbf{A}_{\parallel} - \beta A^0) \tag{3.76}$$

$$\mathbf{A}'_{\perp} = \mathbf{A}_{\perp} \tag{3.77}$$

Combining the last two equations, we get

$$\mathbf{A}' = \mathbf{A} + [(\gamma - 1)/\beta^2]\,(\mathbf{A}\cdot\beta)\beta - \gamma\beta A^0 \tag{3.78}$$

3.4.2 Covariant Form of the Lorentz Condition and Equation of Continuity

The Lorentz gauge condition is given by

$$\frac{1}{c}\frac{\partial\phi}{dt} + \nabla\cdot\mathbf{A} = 0 \tag{3.79}$$

Defining the covariant vector operator

$$\partial_\mu = \frac{\partial}{\partial x^\mu} = \left(\frac{\partial}{\partial x^0}, \nabla\right) \tag{3.80}$$

We can write the contravariant vector operator as

$$\partial^\mu = \frac{\partial}{\partial x_\mu} = \left(\frac{\partial}{\partial x^0}, -\nabla\right) \tag{3.81}$$

where

$$x_\mu = \sum_\nu g_{\mu\nu} x^\nu \tag{3.82}$$

so Eq (3.79) becomes

$$\sum_{\mu=0}^{3} \frac{\partial A^\mu}{\partial x^\mu} = 0$$

or

$$\sum_{\mu=0}^{3} \partial_\mu A^\mu = 0 \tag{3.83}$$

The equation of continuity expresses the conservation of charge and is

$$\nabla\cdot\mathbf{J} + \partial\rho/\partial t = 0 \tag{3.84}$$

Expressed in four-dimensional form, it becomes

$$\sum_{\mu=0}^{3} \frac{\partial j^\mu}{\partial x^\mu} = 0 \tag{3.85}$$

or

$$\sum_{\mu=0}^{3} \partial_\mu j^\mu = 0 \tag{3.86}$$

3.5 TENSOR ALGEBRA. ANTI-SYMMETRIC FOUR-TENSOR

The transformation of a three-vector $\mathbf{A}(A_1, A_2, A_3)$ into a vector $\mathbf{B}(B_1, B_2, B_3)$ is governed by the following transformations:

$$B_1 = T_{11}A_1 + T_{12}A_2 + T_{13}A_3 \quad [3.87(a)]$$

$$B_2 = T_{21}A_1 + T_{22}A_2 + T_{23}A_3 \quad [3.87(b)]$$

$$B_3 = T_{31}A_1 + T_{32}A_2 + T_{33}A_3 \quad [3.87(c)]$$

which may be written in matrix form as

$$\begin{pmatrix} B_1 \\ B_2 \\ B_3 \end{pmatrix} = \begin{pmatrix} T_{11} & T_{12} & T_{13} \\ T_{21} & T_{22} & T_{23} \\ T_{31} & T_{32} & T_{33} \end{pmatrix} \begin{pmatrix} A_1 \\ A_2 \\ A_3 \end{pmatrix} \quad (3.88)$$

or

$$B_j = \sum_k T_{jk} A_k \quad (3.89)$$

where (T_{jk}) is a tensor of second rank and the nine components, T_{jk}, of the linear transformation are called the tensor components.

Let us further determine the law of transformation of the tensor components. If Eq. (3.88) is to be invariant as the coordinates are to be rotated, that is, under orthogonal transformation

$$x'_j = \sum_k \alpha_{jk} x_k \quad (3.90)$$

$$\sum_k \alpha_{ji} \alpha_{jk} = \delta_{ik} \quad (3.91)$$

that implies that T_{jk} must transform to T'_{il} such that

$$B'_i = \sum_l T'_{il} A'_l \quad (3.92)$$

From Eq. (3.89), multiplying by α_{ij}, we get

$$\sum_j \alpha_{ij} B_j = \sum_j \sum_k \alpha_{ij} T_{jk} A_k \quad (3.93)$$

But

$$B'_i = \sum_j \alpha_{ij} B_j ; A_k = \sum_l \alpha_{lk} A'_l$$

hence

$$B'_i = \sum_i \left[\sum_j \sum_k \alpha_{ij} \alpha_{lk} T_{jk} \right] A'_l$$

$$= \sum_i T'_{il} A'_l \quad (3.94)$$

Thus we infer that the components of a tensor of rank 2 transform as

$$T'_{il} = \sum_j \sum_k \alpha_{ij} \alpha_{lk} T_{jk} \quad (3.95)$$

Similarly, the reciprocal transformation is defined as

$$T_{jk} = \sum_j \sum_k \alpha_{ij} \alpha_{lk} T'_{il} \tag{3.96}$$

When $T_{jk} = T_{kj}$, the tensor is said to be symmetric, whereas $T_{jk} = -T_{kj}$ defines, defines an anti-symmetric (or skew-symmetric) tensor. So, $T_{ii} = 0$ for an antisymmetric tensor.

The operations of addition and subtraction of two tensors are defined with respect to their components as

$$R_{jk} = S_{jk} + T_{jk} \qquad [3.97(a)]$$

and

$$R_{jk} = S_{jk} - T_{jk} \qquad [3.97(b)]$$

In a three-dimensional space, an antisymmetric tensor of the second rank reduces to three dependent components and in this sense resembles a vector.
Thus, from Eqs [3.87(a)]–[3.87(c)], when $T_{jk} = -T_{jk}$

$$T = (T_{jk}) = \begin{pmatrix} 0 & -T_{21} & T_{13} \\ T_{21} & 0 & -T_{32} \\ -T_{13} & T_{32} & 0 \end{pmatrix} \tag{3.98}$$

where the vector **T** has the components

$$\begin{aligned} T_1 &= T_{32} \\ T_2 &= T_{13} \\ T_3 &= T_{21} \end{aligned} \tag{3.99}$$

From the definition given by Eqs (3.87), we have

$$\mathbf{B} = \mathbf{T} \times \mathbf{A} \tag{3.100}$$

We construe from Eq. (3.100) that an axial vector is an antisymmetric tensor of the second rank. So the relation $\mathbf{T} = \mathbf{A} \times \mathbf{B}$ written in components is

$$T_{jk} = A_j B_k - A_k B_j = -T_{kj} \tag{3.101}$$

Let us further extend the concept of three-dimensional tensors to a manifold of space–time continuum. Just as four-vectors are defined by the behaviour of their components under general Lorentz transformations, so is also a four-dimensional tensor (four-tensor). But the latter have more components than just four. For example, a four-tensor of rank 2 has 4^2 components $T_{\mu\nu}$, which obey the transformation rule

$$T'_{\rho\lambda} = \sum_\mu \sum_\nu \alpha_{\rho\mu} \alpha_{\lambda\nu} T_{\mu\nu} \tag{3.102}$$

or the inverse transformation

$$T_{\mu\nu} = \sum_{\rho}\sum_{\lambda} \alpha_{\rho\mu}\alpha_{\lambda\nu}T'_{\rho\lambda} \tag{3.103}$$

where α's are Lorentz transformation matrix. In general, a four-tensor of rank n has 4^n components, which are written with n indices and they transform analogously to Eq. (3.102). Four-vectors are four-tensors of rank 1. Many of the basic properties of four-vectors are typical for all four-tensors. For instance, four-tensor (component) equations are form-invariant under general Lorentz transformations; sums, scalar multiples, and scalar derivatives of four tensors are clearly four-tensors.

Example 3.3 Prove that upon rotation of the coordinate system, the components of a vector product of two vectors transform like the components of a four-tensor.

Solution. When the coordinate system is rotated, the components of vectors **A** and **B** are transformed as

$$A_j = \sum_l \alpha_{lj}A'_l$$

$$B_k = \sum_i \alpha_{ik}B'_i$$

Thus,

$$T_{jk} = A_jB_k - A_kB_j$$

$$= \sum_l\sum_i \alpha_{lj}a_{ik}(A'_lB'_i - A'_iB'_l)$$

$$= \sum_l\sum_i \alpha_{lj}\alpha_{ik}T'_{li}$$

which is the correct transformation for tensor components on rotation.

Example 3.4 Prove that $\mathbf{B} = \nabla \times \mathbf{A}$ is an antisymmetric tensor of a second rank.

Solution. The components of $\nabla \times \mathbf{A}$ are

$$T_{jk} = \frac{\partial A_k}{\partial x_j} - \frac{\partial A_j}{\partial x_k} = -T_{kj}$$

and from Eq. (3.102), we have

$$\frac{\partial A'_i}{\partial x'_l} = \sum_j\sum_k \alpha_{ij}\alpha_{lk}\frac{\partial A_j}{\partial x_k}$$

which transform like a tensor, hence constitute a tensor of the second rank.

Example 3.5 Prove that the vector product of two four-vectors is an antisymmetric four-tensor.

Solution. The vector product in four-space is defined as the area of a parallelogram formed by two four-vectors **A** and **B**, drawn from an origin. The components of this product are the projections of the parallelogram on the six coordinate planes, whose areas are

$$T_{\mu\nu} = A_\mu B_\nu - A_\nu B_\mu$$

Thus, the product of two four-vectors is an antisymmetric four-tensor.

3.6 GROUP PROPERTY OF LORENTZ TRANSFORMATION

Lorentz transformations are given by

$$t' = \gamma\left(t - \frac{vx}{c^2}\right) \tag{3.104}$$

$$x' = \gamma(x - vt) \tag{3.105}$$

$$y' = y \tag{3.106}$$

$$z' = z \tag{3.107}$$

where $\gamma = (1 - v^2/c^2)^{-1/2}$.

We can impose on the transformations the condition that they should form a one-parameter group of transformations and so we must require the following:

1. The resultant of two such transformations is also a transformation of the group.
2. The identity transformation belongs to the group.
3. For each transformation, the inverse transformation belongs to the group.
4. The result of two successive transformations should not depend on the order, that is, the transformations are commutative in character.

The Lorentz transformations for all possible relative velocities form a group. Let us apply the group property to Eqs (3.104)–(3.107) and in this way obtain the proof of the law of composition of velocities. For this purpose, we consider two successive Lorentz transformations characterized by $\gamma_1(v_1)$ and $\gamma_2(v_2)$, i.e.,

$$t' = \gamma_1\left(t - \frac{v_1}{c^2}x\right) \tag{3.108}$$

$$x' = \gamma_1(x - v_1 t) \tag{3.109}$$

$$y' = y \tag{3.110}$$

$$z' = z \tag{3.111}$$

and

$$t'' = \gamma_2\left(t' - \frac{v_2}{c^2}x'\right) \tag{3.112}$$

$$x'' = \gamma_2(x' - v_2 t') \tag{3.113}$$

$$y'' = y' \tag{3.114}$$

$$z'' = z' \tag{3.115}$$

These two transformations are equivalent to a single transformation from S to S'', which we can obtain by eliminating t', x', y', z' from Eqs (3.108)—(3.111) and (3.112)—(3.115). Thus,

$$t'' = -\gamma_1 \gamma_2 (v_1/c^2 + v_2/c^2)\, x + \gamma_1\, \gamma_2 (1 + v_1 v_2/c^2) t \tag{3.116}$$

$$x'' = \gamma_1 \gamma_2 (1 + v_1 v_2/c^2)\, x - \gamma_1\, \gamma_2 (v_1 + v_2) t \tag{3.117}$$

$$y'' = y \tag{3.118}$$

$$z'' = z \tag{3.119}$$

On the other hand, expressed in terms of the relative velocity v between S and S'', the transformations must have the form

$$t'' = \gamma t - \gamma \frac{v}{c^2} x \tag{3.120}$$

$$x'' = \gamma x - \gamma v t \tag{3.121}$$

$$y'' = y \tag{3.122}$$

$$z'' = z \tag{3.123}$$

Comparing the coefficients of terms from Eqs (3.116)–(3.119) and (3.120)–(3.123), we have

$$\gamma = \gamma_1 \gamma_2 (1 + v_1 v_2/c^2) \tag{3.124}$$

$$\gamma v = \gamma_1 \gamma_2 (v_1 + v_2) \tag{3.125}$$

Dividing Eq. (3.125) by Eq. (3.124), we obtain

$$v = \frac{v_1 + v_2}{1 + v_1 v_2/c^2} \tag{3.126}$$

which is the correct law of addition of velocities, as borne out by Eq. (2.73). The Lorentz group that we have considered so far did not allow any inclination between the space axes and for the relative velocity to be in an arbitrary direction and contains only one parameter, the relative velocity v. The larger Lorentz group would require six parameters for its specification; three for describing the rotation of the space axes of S' relative to those of S and three other parameters from the components $v_{x'}, v_{y'}, v_z$ that describe the motion of S' relative to S.

The subgroup describing this restricted Lorentz transformation is commutative. If the matrix of one Lorentz transformation is denoted by $\alpha(v_1)$ and of second by $\alpha(v_2)$, then

$$\alpha(v_1)\,\alpha(v_2) = \alpha(v_2)\,\alpha(v_1) = \alpha\left(\frac{v_1 + v_2}{1 + v_1 v_2/c^2}\right) \tag{3.127}$$

Moreover, since v_1 and v_2 enter symmetrically into the expression for v, their order is not relevant to the process of composition, with the consequence that the process of combining two special Lorentz transformations into a composite transformation is commutative.

Furthermore, the identity transformation, $S \rightarrow S'$ is also a special Lorentz transformation, the transformation parameter being equal to zero; this transformation may be visualized as a resultant transformation with v_1 equal and opposite to v_2. Thus, the special Lorentz transformations, for all permissible values of parameter v, constitute a group which is commutative. However, this group is a subgroup of a more general group of Lorentz transformations. The full Lorentz group is not commutative since it involves rotation of axes which are not commutative.

Example 3.6 Deduce Lorentz transformation on the basis of group postulates and isotropy of space.

Solution. Let us consider two inertial frames S and S', the latter moving with velocity **v** with respect to the former and their OX and $O'X'$ axes parallel to each other along the direction of relative velocity.

Furthermore, the origins O and O' coincide at $t = t' = 0$. As the velocity boost is along the x (and x') axis, the perpendicular axes remain unchanged and we will omit them for the sake of brevity. Since the transformation we are seeking connects the two inertial frames, it has to transform a linear motion in the (t, x) plane into a linear motion in (t', x') coordinates. Therefore, it must be a linear transformation. The general form of linear transformation is

$$\begin{pmatrix} t' \\ x' \end{pmatrix} = \begin{pmatrix} \gamma & \delta \\ \beta & \alpha \end{pmatrix} \begin{pmatrix} t \\ x \end{pmatrix} \tag{i}$$

where α, β, γ, and δ are the undetermined functions of relative velocity v.

Let us now consider the motion of the frame S'. Its coordinates in S are $(t, x = vt)$ while in S' are $(t', x' = 0)$. Obviously these two points are connected by our transformation

$$\begin{pmatrix} t' \\ 0 \end{pmatrix} = \begin{pmatrix} \gamma & \delta \\ \beta & \alpha \end{pmatrix} \begin{pmatrix} t \\ vt \end{pmatrix} \tag{ii}$$

which give

$$\beta = -v\alpha \tag{iii}$$

Analogously, considering the motion of the origin of frame S, we get

$$\begin{pmatrix} t' \\ vt' \end{pmatrix} = \begin{pmatrix} \gamma & \delta \\ -\beta & \alpha \end{pmatrix} \begin{pmatrix} t \\ 0 \end{pmatrix} \tag{iv}$$

which gives

$$\beta = -v\gamma \tag{v}$$

Comparing Eqs (iii) and (v), we infer that

$$\alpha = \gamma \tag{vi}$$

Thus the transformation matrix becomes

$$\begin{pmatrix} t' \\ x' \end{pmatrix} = \begin{pmatrix} \gamma & \delta \\ -v\gamma & \gamma \end{pmatrix} \begin{pmatrix} t \\ x \end{pmatrix} \tag{vii}$$

Further, we consider the inverse matrix. This can be done in two ways. Firstly, the inverse matrix is done simply by the usual procedure of inverting it.

$$\begin{pmatrix} t \\ x \end{pmatrix} = \frac{1}{\gamma^2 + v\delta\gamma} \begin{pmatrix} \gamma & -\delta \\ v\gamma & \gamma \end{pmatrix} \begin{pmatrix} t' \\ x' \end{pmatrix} \tag{viii}$$

Alternatively, one can obtain the inverse by substituting $-v$ for v

$$\begin{pmatrix} t \\ x \end{pmatrix} = \begin{pmatrix} \gamma(-v) & \delta(-v) \\ v\gamma(-v) & \gamma(-v) \end{pmatrix} \begin{pmatrix} t' \\ x' \end{pmatrix} \tag{ix}$$

The function is not dependent on the direction of v, since this factor defines the relativistic contraction and time dilation. In an isotropic world, the length contraction as well as the time dilation cannot depend on the direction of v. Therefore, comparing the matrices, Eqs (viii) and (ix), we get

$$\gamma^2 + v\delta\gamma = 1 \tag{x}$$

Finally, composition of two coordinate transformations is also a coordinate transformation. Therefore, the product of two of our matrices will also be a matrix of the same form; in particular the diagonal elements should be equal. Thus, taking the product of two transformation matrices, one with v and the other with v', we get

$$\begin{pmatrix} \gamma(v) & \delta(v) \\ v\gamma(v) & \gamma(v) \end{pmatrix} \begin{pmatrix} \gamma(v') & \delta(v') \\ v'\gamma(v') & \gamma(v') \end{pmatrix} = \begin{pmatrix} \gamma(v)\,\gamma(v') + \delta(v)\,v'\gamma(v') & \gamma(v)\,\delta(v') + \delta(v)\,\gamma(v') \\ v\gamma(v)\,\delta(v') + \gamma(v)\,\delta(v') & v\gamma(v)\,\delta(v') + \gamma(v)\,\gamma(v') \end{pmatrix} \tag{xi}$$

which gives

$$\frac{v\gamma(v)}{\delta(v)} = \frac{v'\gamma(v')}{\delta(v')} \tag{xii}$$

Since this is valid for any arbitrary v and v', this combination function must be a universal constant, same for all inertial frames. Defining this constant as

$$\frac{v\gamma(v)}{\delta(v)} = -c^2 \tag{xiii}$$

where c has the dimensions of velocity. We get from Eq. (x)

$$\gamma = \frac{1}{\sqrt{1 - v^2/c^2}} \tag{xiv}$$

Thus the transformation matrix becomes

$$\begin{pmatrix} t' \\ x' \end{pmatrix} = \frac{1}{\sqrt{1 - v^2/c^2}} \begin{pmatrix} 1 & -v/c^2 \\ -v & 1 \end{pmatrix} \begin{pmatrix} t \\ x \end{pmatrix} \tag{xv}$$

It may be remarked that c^2 cannot be negative, since it will transform time into spatial coordinates and vice versa. This is undesirable since time can run only in the positive direction whereas spatial coordinates can run in both the directions. If $c^2 > 0$, it can be the highest achievable velocity. If it is infinitely large, it gives Galilean transformation and Euclidean world with absolute time; or it can be finite, which gives Lorentz transformation and Minkowski space of special relativity. The experiment tells that it is finite, $c = 2.99792458 \times 10^8$ m/s.

The general Lorentz transformation is given by

$$\mathcal{L} = \begin{vmatrix} \alpha_{00} & \alpha_{01} & \alpha_{02} & \alpha_{03} \\ \alpha_{10} & \alpha_{11} & \alpha_{12} & \alpha_{13} \\ \alpha_{20} & \alpha_{21} & \alpha_{22} & \alpha_{23} \\ \alpha_{30} & \alpha_{31} & \alpha_{32} & \alpha_{33} \end{vmatrix} \tag{3.128}$$

So far we had implied by special Lorentz transformation that transformation that leaves the interval $ds^2 = c^2\,dt^2 - dx^2 - dy^2 - dz^2$ invariant. But the mere invariance of ds^2 gives a group that is wider than the group that we had employed so far. For example, the transformation $t' = -t, x' = x, y' = y, z' = z$ leaves ds^2 unchanged but it involves time reversal, that is, the interchange of past and future. This can be avoided if we impose the restriction that $\alpha_{00} > 0$ in all elements of the group. Another example of a transformation that leaves ds^2 unchanged but implies space inversion is $t' = t, x' = -x, y' = -y, z' = -z$.

This transformation converts a right-handed system S into a left-handed system S' or vice versa. This space inversion can be avoided by the additional requirement that det $\mathcal{L} = +1$.

The determinant for any Lorentz transformation can be only ± 1. Since the identical transformation has a determinant $+1$, the restricted group can consist of those transformation matrices with det $\mathcal{L} = +1$. Lorentz transformations corresponding to a determinant -1 do not form a group because the product of an even number of matrices yields a matrix with a determinant $+1$.

Consider the general Lorentz transformations given by Eq. (3.24). Since it is an orthogonal transformation, we get the following result from the condition of orthogonality given by Eq. [3.24(a)]:

$$\alpha_{00}^2 + \alpha_{10}^2 + \alpha_{20}^2 + \alpha_{30}^2 = 1$$

or

$$\alpha_{00}^2 = 1 - \left(\alpha_{10}^2 + \alpha_{20}^2 + \alpha_{30}^2\right) \tag{3.129}$$

Since $\alpha_{10}^2 + \alpha_{20}^2 + \alpha_{30}^2 < 0$, for when $x_0 = ict$, the coefficients $\alpha_{10} + \alpha_{20} + \alpha_{30}$ are purely imaginary,

or

$$|\alpha_{00}| \geq 1 \tag{3.130}$$

Thus there is a gap of at least 2 in the value of α_{00} between transformations with $\alpha_{00} > 0$ and $\alpha_{00} < 0$. This gap cannot be abridged by continuous steps.

The general Lorentz group can be subdivided into four sets of transformations, each of which is continuous within itself but not among each other. These subdivisions are

	I	II	III	VI
det $\mathcal{L}$	+1	−1	−1	+1
α_{00}	>0	>0	<0	<0

The first set constitutes the restricted Lorentz group (the special Lorentz transformations which correspond to rotations in Minkowski space). The second set consists of Lorentz transformations with space inversion and it does not form a group. Sets I and II form a group (orthochronous Lorentz group) that consists of all the transformations which do not involve time reversal. The set III consists of transformations which involve time reversal and space reflection and these transformations do not form a group. Set IV consists of transformations which involve time reversal. The transformation with det $\mathcal{L} = +1$ are called *proper* or *unimodular* transformations such as I and IV; and those with det $\mathcal{L} = -1$, improper transformations such as II and III. The proper transformations I and IV also form a group.

Reference

1. H. Minkowski, *Phys. Z.* (1909) *10*, 104; cf. also H. Poincaré, *Rend. Palermo* (1906) *21*, 129. On imaginary time in special relativity, W. Band, *Am. J. Phys.* (1938) *6*, 323.

Problems

3.1 Prove that an orthogonal transformation

$$x'_\mu = \sum_{\nu=0}^{3} \alpha_{\mu\nu} x_\nu$$

leaves the sum of the squares of the coordinates invariant.

3.2 An infinitesimal homogeneous linear transformation $x_\mu \to x'_\mu$ is defined by the relation $x'_\mu = x_\mu + \varepsilon_{\mu\nu} x_\nu = (\delta_{\mu\nu} + \varepsilon_{\mu\nu}) x_\nu$, where the $\varepsilon_{\mu\nu}$ are infinitesimal quantities. Show that the orthogonality condition in the case of infinitesimal Lorentz transformation reduces to the condition that $\varepsilon_{\mu\nu}$ be antisymmetric in μ and ν.

3.3 Show that Lorentz transformation can be characterized by the relation $\mathcal{L}^T = \mathcal{L}^{-1}$.

3.4 Find $\mathcal{L}^{-1}(v)$ and show that $\mathcal{L}^{-1}(v) = \mathcal{L}(-v)$.

3.5 Show algebraically that two successive Lorentz transformations at right angles (v_1 in the x-direction, v_2 in the y-direction) do not commute. Further show that in whatever order they are applied, the result is not the same as a single transformation with $v = \mathbf{i}_1 v_1 + \mathbf{i}_2 v_2$.

3.6 Given that if $\theta = v/c$ the Lorentz transformations are

$$x = x' \cosh\theta + ct' \sinh\theta$$
$$ct = x' \sinh\theta + ct' \cosh\theta$$

3.7 The Lorentz transformation in Problem 3.6 can be written in the matrix form as

$$\begin{pmatrix} x \\ ct \end{pmatrix} = \begin{pmatrix} \cosh\theta & \cosh\theta \\ \sinh\theta & \sinh\theta \end{pmatrix} \begin{pmatrix} x' \\ ct' \end{pmatrix}$$

Show that the result of applying Lorentz transformation with $\tanh\theta_1 = v_1/c$ followed by another Lorentz transformation with $\tanh\theta_2 = \frac{v_2}{c}$, is a Lorentz transformation with $\tanh(\theta_1 + \theta_2) = \frac{v}{c}$, where $v = \dfrac{v_1 + v_2}{1 + v_1 v_2/c^2}$.

3.8 Show that the energy–momentum relationship $\mathcal{E}^2 = p^2c^2 + m^2c^4$ from $\mathcal{E} = \gamma mc^2$ and $p = \gamma mv$.

3.9 A particle of mass m is moving in the $+x$-direction with speed v and has momentum p and energy $\mathcal{E}$ in the frame S. If another inertial frame S' is moving with speed v in the standard way, determine the momentum p' and energy $\mathcal{E}'$ observed in S' and thereby show that $\mathcal{E}^2 - p^2c^2$ is an invariant quantity, i.e.,

$$\mathcal{E}'^2 - p'^2c^2 = \mathcal{E}^2 - p^2c^2$$

3.10 The four-momentum of a particle with non-zero rest mass is $p^\mu = mu^\mu$ where u^μ is the four-velocity. Show that the squared magnitude of the four-momentum is $-m^2c^2$.

4

Relativistic Mechanics of a Particle. Collisions and Conservation Laws

The theory of relativity does not permit the existence of rigid bodies and the argument goes as follows. A rigid body is defined as a body which is non-deformable under all conditions of rest and motion. Suppose a rigid body is set into motion by applying a force from outside. The particle at the point of application will acquire the motion imparted instantly whereas the rest of the particles which constitute the body will be set into motion eventually, depending upon the velocity of propagation of the impressed interaction. Since the latter has finite velocity, there is bound to be a time lag in the different parts to acquire the motion and this is sure to cause deformation of the rigid body. The existence of rigid bodies is thus in contradistinction to the principle of relativity.

An elementary particle which has no finite dimensions and is represented by a geometrical point is the only entity which is capable of existence at relativistic speeds.[1–4]

4.1 RELATIVISTIC LAGRANGIAN FOR A FREE PARTICLE

We will be determining the form of Lagrangian for a free particle by getting an expression for its action and comparing the action integral, expressed in terms of time, with

$$A = \int_{t_1}^{t_2} \mathcal{L} dt \tag{4.1}$$

where $\mathcal{L}$ is the Lagrangian.

We will like to derive an expression for action subject to the following considerations:

1. Action must be invariant to Lorentz transformation, so it must be a scalar.
2. Lagrangian for a free particle must be a function of only the absolute velocity, rather than velocity vector **u**, since in free space there is no preferred direction in which it could be turned.
3. The integrand must be a differential of the first order.

The simplest kind of scalar that one could ascribe to a free particle is $|\alpha\, ds|$ where ds is the interval. So the action for a free particle is

$$\begin{aligned} A &= \int \alpha\, ds \\ &= \int \alpha c \sqrt{1 - u^2/c^2}\, dt \end{aligned} \tag{4.2}$$

where α is a constant characterizing the particle. The limits of the integral refer to two definite times on a world line.

Thus the Lagrangian of the particle is

$$\mathcal{L} = \alpha c\sqrt{1 - u^2/c^2}$$

$$\approx \alpha c - \alpha u^2/2c \qquad \text{for } u << c \tag{4.3}$$

The first term is a constant and so can be omitted since it does not appear in Lagrange's equations. Furthermore, the addition of any total time derivative of function of coordinates and time to $\mathcal{L}$, does not alter the equations.

Comparing $-\frac{1}{2}(\alpha u^2/c)$ with $\frac{1}{2}mu^2$ in the Lagrangian for a free particle in Newtonian mechanics, we infer that

$$\alpha = -mc$$

Thus finally, the Lagrangian for a free particle

$$\mathcal{L} = -mc^2\sqrt{1 - u^2/c^2} \tag{4.4}$$

where m is the mass of the particle measured in a coordinate system in which the particle is at rest or very nearly so.

4.2 ENERGY AND MOMENTUM OF A FREE PARTICLE

4.2.1 Three-dimensional Form

Momentum is defined as the vector

$$\mathbf{p} = \frac{\partial \mathcal{L}}{\partial \mathbf{u}} \tag{4.5}$$

Hence, momentum for a free particle, from Eq. (4.4), is

$$\mathbf{p} = \frac{\partial}{\partial \mathbf{u}}\left(-mc^2\sqrt{1 - u^2/c^2}\right)$$

$$= \frac{m\mathbf{u}}{\sqrt{1 - u^2/c^2}} \tag{4.6}$$

Now when $\mathbf{u} << c$, the momentum $\mathbf{p} = m\mathbf{u}$, as is the case classically. The maximum value of $\mathbf{u}$ is c, since if $u > c$, the momentum of the particle becomes imaginary which is meaningless.

The energy $\mathcal{E}$ is given by

$$\mathcal{E} = \mathbf{p}\cdot\mathbf{u} - \mathcal{L}$$

$$= \frac{mu^2}{\sqrt{1 - u^2/c^2}} + mc^2\sqrt{1 - u^2/c^2}$$

$$= \frac{mc^2}{\sqrt{1 - u^2/c^2}} \tag{4.7}$$

When the velocity of the particle u becomes zero, the corresponding energy is denoted by $\mathcal{E}_0 = mc^2$. This is called the *rest energy.*

Let us evaluate the energy of a particle in the non-relativistic approximation from Eq. (4.7). Thus expanding the denominator

$$\mathcal{E} = mc^2\left(1 - u^2/c^2\right)^{-1/2}$$

$$= mc^2 + \frac{1}{2}mu^2 + \frac{3}{8}m\left(\frac{u^2}{c}\right)^2 + \ldots \tag{4.8}$$

Comparing Eqs (4.6) and (4.8), we get the relation between the momentum and energy of a particle

$$\mathbf{p} = \frac{\mathcal{E}}{c^2}\mathbf{u} \tag{4.9}$$

when $u = c$, which is the case for a photon,

$$p = \mathcal{E}/c$$

or

$$\mathcal{E} = cp \tag{4.10}$$

When the particle moves in a field of potential ϕ, then its energy is

$$\mathcal{E} = mc^2/\sqrt{1 - u^2/c^2} + \phi \tag{4.11}$$

4.2.2 Four-dimensional Formulation. Energy–Momentum Four-Vector

From Eq. (4.2), we have

$$A = -mc\int ds \tag{4.12}$$

Therefore,

$$\delta A = -mc\delta\int ds \tag{4.13}$$

Substituting $ds = \sqrt{dx_\mu dx^\mu}$ in the above equation

$$\delta A = -\frac{1}{2}mc\int_a^b \frac{dx_\mu d\delta x^\mu + dx^\mu\, d\delta x_\mu}{ds}$$

$$\delta A = -mc\int_a^b \frac{dx_\mu \delta dx^\mu}{ds} = -mc\int_a^b u_\mu\, \delta dx^\mu$$

where $u_\mu = \frac{dx_\mu}{ds}$, is the four-velocity. Integrating by parts, we get

$$\delta A = -mcu_\mu\, \delta x^\mu \Big|_a^b + mc\int_a^b \delta x^\mu \frac{du_\mu}{ds}\, ds$$

As usual, to get the equation of motion, we compare different trajectories between two fixed points and the actual trajectory is the one corresponding to $\delta A = 0$. Thus, the first term on the right-hand side vanishes since the limits $(\delta x^\mu)_a = (\delta x^\mu)_b = 0$. Thus we obtain the equations

$$\frac{du_\mu}{ds} = 0 \tag{4.14}$$

Denoting $\omega_\mu = du_\mu/ds$, the four-acceleration, the equation of motion of a free particle becomes

$$\omega_\mu = 0 \tag{4.15}$$

This implies that a free particle has a constant velocity in space–time continuum.

The variation of action δA as a function of coordinates is determined by keeping one limit fixed and considering the other as variable and admitting those trajectories for which $\omega_\mu = 0$. Then,

$$\delta A = \left[-mcu_\mu \, \delta x^\mu \,|_a^b + \ldots\right]$$

or

$$\frac{\delta A}{\delta x^\mu} = -mcu_\mu \tag{4.16}$$

The four-vector

$$p_\mu = -\frac{\partial A}{\partial x^\mu} \tag{4.17}$$

is called the *momentum four-vector*. From mechanics, we know that the derivatives $\partial A/\partial x$, $\partial A/\partial y$, $\partial A/\partial z$ are the three components of the momentum vector $\mathbf{p}$ of the particle, while the derivative $-\partial A/\partial t$ is the energy of the particle $\mathcal{E}$. Thus the covariant components of the four-momentum are $p_\mu = (\mathcal{E}/c, -\mathbf{p})$, while the contravariant components are

$$p^\mu = (\mathcal{E}/c, \mathbf{p}) \tag{4.18}$$

which in component form may be written as follows:

$$p^\mu = \left(\frac{mc}{\sqrt{1 - u^2/c^2}}, \frac{m\mathbf{u}}{\sqrt{1 - u^2/c^2}}\right) \tag{4.19}$$

In relativistic mechanics, the energy and momentum are the components of a single four-vector p^μ. Making use of the general transformation for components of a four-vector, we show that the corresponding transformations for the momentum four-vector are

$$\mathcal{E} = \frac{\mathcal{E}' + up_x'}{\sqrt{1 - u^2/c^2}} \tag{4.20(a)}$$

$$p_x = \frac{p'_x + u\mathcal{E}'/c^2}{\sqrt{1 - u^2/c^2}} \qquad [4.20(b)]$$

$$p_y = p'_y \qquad [4.20(c)]$$

$$p_z = p'_z \qquad [4.20(d)]$$

where $\mathcal{E}, p_x, p_y, p_z$ refer to the unprimed system S and $\mathcal{E}', p'_x, p'_y, p'_z$ to the primed system S' moving with velocity u relative to S, the axis OX being parallel to axis $O'X'$.

The Lorentz invariant of the four-momentum p^μ is

$$\sum_{\mu=0}^{3} p^\mu p_\mu = \mathcal{E}^2/c^2 - p^2 \qquad (4.21)$$

Alternatively we have

$$\sum_{\mu=0}^{3} p^\mu p_\mu = \sum_{\mu=0}^{3} (mu^\mu)(mu_\mu) = m^2 \sum_{\mu=0}^{3} u^\mu u_\mu = m^2c^2$$

and hence we have

$$\mathcal{E}^2/c^2 - p^2 = m^2c^2 \qquad (4.22)$$

This is called energy–momentum invariant, since the square of energy–momentum four-vector, like any four-vector, is a scalar.

Thus,

$$\mathcal{E}^2 = p^2c^2 + m^2c^4 \qquad (4.23)$$

or

$$\mathcal{E} = \pm\sqrt{m^2c^4 + c^2p^2}$$

Thus the fact that a free particle may have a positive or negative energy is a consequence of relativity theory. Pre-quantum mechanics physics could not give any explanation for this conclusion. Dirac equation admits the existence of states with negative energy but these are not observed in actual experiment.

The energy expressed in terms of coordinates and momentum is called Hamiltonian and thus the relativistic Hamiltonian of a free particle is

$$\mathcal{H} = c\sqrt{p^2 + m^2c^2} \qquad (4.24)$$

However, if the particle is in some external field, Eq. (4.24) becomes

$$\mathcal{H} = c\sqrt{p^2 + m^2c^2} + U \tag{4.25}$$

where U is the potential energy of the particle in the external field.

Example 4.1 Prove the Lorentz invariance of the following expressions:

(i) $\dfrac{dp_x\, dp_y\, dp_z}{\mathcal{E}}$

(ii) $pd\mathcal{E}d\omega$

where $d\omega$ is the element of solid angle around the direction of vector **p**.

Solution. (i) For the expression to be Lorentz invariant, we must have

$$\frac{dp_x\, dp_y\, dp_z}{\mathcal{E}} = \frac{dp'_x\, dp'_y\, dp'_z}{\mathcal{E}'}$$

But,

$$\mathcal{E} = \frac{\mathcal{E}' + vp'_x}{\sqrt{1 - v^2/c^2}}$$

$$p_x = \frac{p'_x + v\mathcal{E}'/c^2}{\sqrt{1 - v^2/c^2}}$$

$$p_y = p'_y$$

$$p_z = p'_z$$

Hence, in view of these transformations, we have to show that

$$\frac{dp_x}{\mathcal{E}} = \frac{dp'_x}{\mathcal{E}'}$$

$$= \frac{dp_x - \frac{v}{c^2}\, d\mathcal{E}}{\mathcal{E} - vp_x}$$

or

$$1 - \frac{vd\mathcal{E}}{c^2\, dp_x} = 1 - \frac{vp_x}{\mathcal{E}}$$

or

$$\mathcal{E}\frac{d\mathcal{E}}{dp_x} = c^2\, p_x$$

But

$$\mathcal{E}^2 = c^2p^2 + m^2c^4$$

$$= c^2\left(p_x^2 + p_y^2 + p_z^2\right) + m^2c^4$$

Therefore,

$$\frac{\mathcal{E}d\mathcal{E}}{dp_x} = c^2 p_x$$

Hence $\frac{(dp_x\, dp_y\, dp_z)}{\mathcal{E}}$ is Lorentz invariant.

(ii) Further let us consider the expression $pd\mathcal{E}d\omega$

We know that

$$\mathcal{E}^2 = c^2p^2 + m^2c^4$$

and

$$pdp = \frac{\mathcal{E}d\mathcal{E}}{c^2}$$

Hence

$$\frac{p^2\, dpd\omega}{\mathcal{E}} = \frac{pd\mathcal{E}d\omega}{c^2}$$

But the left-hand side is invariant under Lorentz transformation, hence $pd\mathcal{E}d\omega$ is also Lorentz invariant.

4.3 THE CONSERVATION OF MOMENTUM. VARIATION OF MASS WITH VELOCITY

It is of paramount interest to carry over into relativistic mechanics as much of classical mechanics as possible. The law of addition of velocities as derived from Lorentz transformation is different from the one which one would expect from Galilean transformation. Hence, if we intend to preserve the classical law of conservation of momentum, then we are forced to modify the classical concept of mass.

In Newtonian mechanics, a particle of mass m moving with velocity $\mathbf{u}$ has linear momentum $\mathbf{p}$, which is given by

$$\mathbf{p} = m\mathbf{u} \tag{4.26}$$

where $\mathbf{u} = d\mathbf{r}/dt$. The vector $\mathbf{u}$ does not constitute the last three components of a four-vector $dx_\mu = (cdt, d\mathbf{r})$, since while $d\mathbf{r}$ represents the last three components of it, dt is not an invariant. We know from Eq. (4.17), that

$$p_\mu = mcu_\mu \tag{4.27}$$

where m is the inertial mass of the particle, i.e., mass measured in the proper frame of the particle. Defined in this way, $p_\mu p^\mu$ is Lorentz invariant.

Consider two particles of masses m_1 and m_2 and having velocities u_1 and u_2 interacting at a point. Then the classical law of conservation of momentum states that

$$m_1 u_1 + m_2 u_2 = \text{constant} \tag{4.28}$$

If the law is to be independent of the choice of inertial frame, then it must take the form

$$p_{\mu 1} + p_{\mu 2} = \text{constant} \tag{4.29}$$

where $p_{\mu 1}$ is the four-momentum of the particle number 1 and $p_{\mu 2}$ that of particle number 2. The components of p_μ are

$$\left(\frac{mc}{\sqrt{1-u^2/c^2}}, \frac{m\mathbf{u}}{\sqrt{1-u^2/c^2}}\right) \tag{4.30}$$

where $\mathbf{u}$ is the velocity in the given frame. The three spatial components have the form

$$\mathbf{p} = m_r \mathbf{u}$$

where

$$m_r = \frac{m}{\sqrt{1-u^2/c^2}} \tag{4.31}$$

Thus we infer that the law of variation of mass with velocity follows as an immediate consequence of formulating the law of conservation of momentum in a covariant manner. The variation of mass of a particle with velocity also follows from the following *thought* experiment.

Consider an elastic head-on collision between two similar point particles. Denote as S' the centre of mass frame in which the collision is symmetrical, i.e., they appear with equal velocities and after the collision move back over their original paths with the respective velocities $-\mathbf{u}'$ and $+\mathbf{u}'$ of the same magnitude as before but reversed in direction (Figure 4.1), and $\mathbf{v}$ is the velocity of S' w.r.t. S.

Assuming that momentum and mass are to be conserved in any frame, so in S

$$m_1 + m_2 = M \tag{4.32}$$

$$m_1 u_1 + m_2 u_2 = Mv \tag{4.33}$$

where M is the combined mass during the collision. From the law of composition of velocities, we have

$$u_1 = \frac{u' + v}{1 + u'v/c^2} \tag{4.34}$$

$$u_2 = \frac{-u' + v}{1 - uv/c^2} \tag{4.35}$$

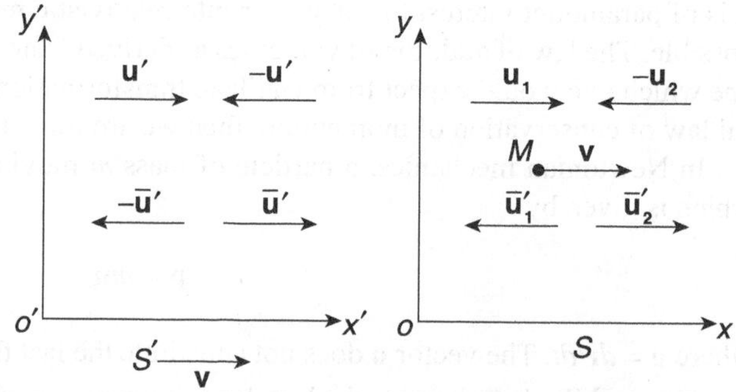

Figure 4.1 *Collision between two similar mass points as observed in S and S′, v is the velocity of S′ w.r.t. S. The bars indicate the quantities after the collision.*

Eliminating M from Eqs (4.32) and (4.33) and employing Eqs (4.34) and (4.35), we get

$$\frac{m_1}{m_2} = \frac{1 + u'v/c^2}{1 - u'v/c^2} \tag{4.36}$$

Let us evaluate the expression $(1 + u'v/c^2)/(1 - u'v/c^2)$.
Making use of the result of Example 2.8, we get

$$1 + \frac{vu'_x}{c^2} = \left[\frac{(1 - v^2/c^2)(1 - u'^2/c^2)}{1 - u^2/c^2}\right]^{1/2} \tag{4.37}$$

For particle 1, one has

$$\text{in } S'\text{: } u'_x = u'\text{; and } u' = u'$$

and

$$\text{in } S\text{: } u = u_1$$

Substituting these values in the above result, one gets

$$1 + vu'/c^2 = \left[\frac{(1 - v^2/c^2)(1 - u'^2/c^2)}{1 - u_1^2/c^2}\right]^{1/2}$$

For particle 2, one gets

$$\text{in } S'\text{: } u'_x = -u'$$

and

$$\text{in } S\text{: } u = u_2$$

Substituting these values in the same result again, one has

$$1 - vu'/c^2 = \left(\frac{(1 - v^2/c^2)(1 - u'^2/c^2)}{1 - u_2^2/c^2}\right)^{1/2} \tag{4.39}$$

Dividing Eq. (4.38) by Eq. (4.39), one gets

$$\frac{1 + vu'/c^2}{1 - vu'/c^2} = \frac{\sqrt{1 - u_2^2/c^2}}{\sqrt{1 - u_1^2/c^2}} \tag{4.40}$$

Therefore one has

$$\frac{m_1}{m_2} = \frac{\sqrt{1 - u_2^2/c^2}}{\sqrt{1 - u_1^2/c^2}} \tag{4.41}$$

Hence if we have to preserve the law of conversation of momentum in all frames, for relativistic mass moving with speed u, we must have

$$m_r = \frac{m}{\sqrt{1 - u^2/c^2}} \tag{4.42}$$

where m is the proper mass of the particle. Expanding it by Binomial theorem, we have

$$m_r = m\left[1 + 1/2\left(\frac{u}{c}\right)^2 + 3/8\left(\frac{u}{c}\right)^4 + 15/48\left(\frac{u}{c}\right)^6 + \cdots\right] \tag{4.43}$$

Such a variation of mass with velocity was indeed observed by Kaufmann[5] in experiments designed to measure the e/m ratio of electrons of radioactive origin. This has also been verified to a very high degree of accuracy; since the behaviour of high-energy accelerators designed using Eq. (4.42) confirms the prediction.

Example 4.2 Consider two particles of masses m_1 and m_2 and having velocities $\mathbf{u}_1$ and $\mathbf{u}_2$ in frame S. Assuming that mass and momentum are to be conserved in any frame, so in S

$$m_1 + m_2 = M$$

$$m_1 u_1 + m_2 u_2 = Mv$$

where M is the combined mass during collision. Deduce the velocity dependence of mass, using the law of composition of velocities.

Solution. S and S' are two frames of reference and S' is moving with velocity v along OX. Let m_1 be the mass of a particle moving with velocity u_1 in system S and let the corresponding mass and velocity be m_1' and u_1' as measured by the observer in S'.
Denote

$$\gamma = \frac{1}{\sqrt{1 - v^2/c^2}},\quad \gamma_1 = \frac{1}{\sqrt{1 - u_1^2/c^2}},\quad \gamma_1' = \frac{1}{\sqrt{1 - u_1'/c^2}} \tag{i}$$

According to the law of addition of velocities, we have

$$u_1 = \frac{u_1' + v}{1 + vu_1'/c^2}$$

and

$$u_1' = \frac{u_1 - v}{1 - vu_1/c^2} \tag{ii}$$

Therefore,

$$\gamma_1' u_1' = \frac{u_1 - v}{\left(1 - u_1'^2/c^2\right)^{1/2}\left(1 - vu_1/c^2\right)} \tag{iii}$$

Let us evaluate

$$1 - u_1'/c^2 = 1 - \frac{(u_1 - v)^2}{c^2\left(1 - vu_1/c^2\right)^2}$$

$$c^2\left(1 - u_1'^2/c^2\right)\left(1 - vu_1/c^2\right)^2 = c^2\left(1 - vu_1/c^2\right)^2 - (u_1 - v)^2$$

$$= c^2\left(1 + \frac{v^2u_1^2}{c^4} - \frac{2u_1v}{c^2}\right) - (u_1^2 + v^2 - 2u_1v)$$

$$= c^2\left(1 - \frac{u_1^2 + v^2}{c^2} + \frac{u_1^2\,v^2}{c^4}\right)$$

which gives

$$\left(1 - \frac{u_1'^2}{c^2}\right)\left(1 - \frac{vu_1}{c^2}\right)^2 = 1 - \frac{u_1^2 + v^2}{c^2} + \frac{u_1^2\,v^2}{c^4}$$

Substituting this result into Eq. (iii), we get

$$\gamma_1'\,u_1' = \frac{u_1 - v}{\left(1 - \dfrac{u_1^2 + v^2}{c^2} + \dfrac{u_1^2\,v^2}{c^4}\right)^{1/2}} \tag{iv}$$

$$= \frac{u_1 - v}{\left[\left(1 - u_1^2/c^2\right)\left(1 - v^2/c^2\right)\right]^{1/2}}$$

$$= \gamma\,\gamma_1\,(u_1 - v)$$

in view of Eq. (i). Therefore,

$$\gamma_1'\,u_1' = \gamma\,\gamma_1\,(u_1 - v)$$

or

$$\frac{\gamma_1'\,u_1'}{\gamma_1} = \gamma\,(u_1 - v) \tag{v}$$

Assume that there are a number of particles in S that are moving along x-axis, and their masses and momenta are invariant so that

$$\sum m_1 = \text{constant} \qquad \text{[vi(a)]}$$

and

$$\sum m_1\,u_1 = \text{constant} \qquad \text{[vi(b)]}$$

All the particles have the same γ and v, so from Eqs [vi(a)] and [vi(b)], we get

$$\sum m_1\,\gamma\,v = \text{constant} \qquad \text{[vii(a)]}$$

and

$$\sum m_1\,u_1\gamma = \text{constant} \qquad \text{[vii(b)]}$$

Subtracting one from the other, we get

$$\Sigma m_1 \gamma(u_1 - v) = \text{constant} \tag{viii}$$

Rewriting Eq. (viii) in the light of Eq. (v), we have

$$\sum m_1 \frac{\gamma_1' u_1'}{\gamma_1} = \text{constant} \tag{ix}$$

According to the law of conservation of momentum in System S', we have

$$\sum m_1' u_1' = \text{constant} \tag{x}$$

Comparing Eqs (ix) and (x), we get

$$\frac{m_1 \gamma_1'}{\gamma_1} = m_1'$$

or

$$\frac{m_1}{\gamma_1} = \frac{m_1'}{\gamma_1'} = \text{absolute constant, say } m_0$$

Therefore,

$$m_1 = m_0/(1 - u_1^2/c^2)^{1/2}$$

and

$$m_1' = m_0/(1 - u_1'^2/c^2)^{1/2}$$

Thus if a particle of mass m in system S is moving with velocity u relative to S, then

$$m' = m_0/(1 - u^2/c^2)^{1/2}$$

If $\mathbf{u} = 0$, then $m = m_0$ and m_0 is called the mass of the particle at rest or the *proper* mass of the particle.

4.3.1 Transformation of Mass

S and S' are two inertial systems and S' is moving with velocity v along OX. Let m and m' be the masses of a body moving with velocity u and u' as measured by the observers in S and S', respectively. Now

$$m = \frac{m_0}{\sqrt{1 - u^2/c^2}}, \quad m' = \frac{m_0}{\sqrt{1 - u'^2/c^2}} \tag{4.44}$$

$$u^2 = u_x^2 + u_y^2 + u_z^2; \; u'^2 = u_x'^2 + u_y'^2 + u_z'^2 \tag{4.45}$$

Furthermore, from the law of composition of velocities, we have

$$u'_x = \frac{u_x - v}{1 - \frac{vu_x}{c^2}} \tag{4.46}$$

$$u'_y = \frac{u_y\sqrt{1 - u^2/c^2}}{1 - \frac{vu_x}{c^2}} \tag{4.47}$$

$$u'_z = \frac{u_z\sqrt{1 - u^2/c^2}}{1 - \frac{vu_x}{c^2}} \tag{4.48}$$

From Eq. (4.44), we have

$$\frac{m}{m'} = \frac{\sqrt{1 - u'^2/c^2}}{\sqrt{1 - u^2/c^2}} \tag{4.49}$$

Let us evaluate the expression

$$1 - \frac{u'^2}{c^2} = 1 - \frac{1}{c^2}\left(u_x'^2 + u_y'^2 + u_z'^2\right)$$

$$= 1 - \frac{1}{c^2(1 - vu_x/c^2)^2}\left[(u_x - v)^2 + u_y^2\left(1 - v^2/c^2\right) + u_z^2\left(1 - v^2/c^2\right)\right]$$

$$= \frac{1}{c^2(1 - vu_x/c^2)^2}\left[c^2\left(1 - vu_x/c^2\right)^2 - \left\{(u_x - v)^2 + u_y^2\left(1 - v^2/c^2\right) + u_z^2\left(1 - v^2/c^2\right)\right\}\right]$$

$$= \frac{1}{c^2(1 - vu_x/c^2)^2}\left[c^2\left(1 - v^2/c^2\right) + v^2u^2/c^2 - u^2\right]$$

$$= \frac{1}{c^2(1 - vu_x/c^2)^2}c^2\left(1 - u^2/c^2\right)\left(1 - v^2/c^2\right)$$

Taking square root, we get

$$\sqrt{1 - u'^2/c^2} = \frac{\sqrt{1 - u^2/c^2}\sqrt{1 - v^2/c^2}}{(1 - vu_x/c^2)} \tag{4.50}$$

This is the transformation of Lorentz contraction factor.
Employing Eq. (4.49), we have

$$\frac{m}{m'} = \frac{\sqrt{1 - v^2/c^2}}{(1 - vu_x/c^2)}$$

or

$$m' = \frac{m(1 - vu_x/c^2)}{\sqrt{1 - v^2/c^2}} \tag{4.51}$$

This is the required transformation for mass. When $u_x = 0$, then

$$m' = \frac{m}{\sqrt{1 - v^2/c^2}} \tag{4.52}$$

4.4 FOUR-DIMENSIONAL VELOCITY AND ACCELERATION

A four-vector is a set of four quantities a^0, a^1, a^2, a^3 which transforms from one system to another according to the same rules as the components of a four-position vector x^0, x^1, x^2, x^3. Thus we have the transformation formulae

$$a'^0 = \frac{a^0 - \beta a^1}{\sqrt{1 - \beta^2}} \qquad [4.53(a)]$$

$$a'^1 = \frac{-\beta a^0 + a^1}{\sqrt{1 - \beta^2}} \qquad [4.53(b)]$$

$$a'^2 = a^2 \qquad [4.53(c)]$$

$$a'^3 = a^3 \qquad [4.53(d)]$$

and the inverse transformations are

$$a^0 = \frac{a'^0 + \beta a'^1}{\sqrt{1 - \beta^2}} \qquad [4.54(a)]$$

$$a^1 = \frac{\beta a'^0 + a'^1}{\sqrt{1 - \beta^2}}$$

$$a^2 = a'^2 \qquad [4.54(c)]$$

$$a^3 = a'^3 \qquad [4.54(d)]$$

The set of four quantities dx^μ/dt does not behave like a four-vector since dt is not invariant. However, we can constitute an invariant from dt through the proper time $d\tau = ds/c$. Thus the quantities

$$u^\mu = \frac{dx^\mu}{d\tau} = c\frac{dx^\mu}{ds} \tag{4.55}$$

constitute a four-vector since $d\tau$ is invariant (scalar). u^μ is called the four-velocity of the particle.

Identically, we define another four-vector called four-acceleration of a particle as

$$\omega^\mu = \frac{d^2x^\mu}{d\tau^2} = \frac{du^\mu}{d\tau} = c\frac{du^\mu}{ds} \tag{4.56}$$

The components of the four-velocity are

$$u^0 = \frac{dx^0}{d\tau} = \frac{cdt}{dt\sqrt{1-\beta^2}} = \frac{c}{\sqrt{1-\beta^2}} \tag{4.57(a)}$$

$$u^1 = \frac{dx^1}{d\tau} = \frac{dx}{dt\sqrt{1-\beta^2}} = \frac{u_x}{\sqrt{1-\beta^2}} \tag{4.57(b)}$$

$$u^2 = \frac{dx^2}{d\tau} = \frac{dy}{dt\sqrt{1-\beta^2}} = \frac{u_y}{\sqrt{1-\beta^2}} \tag{4.57(c)}$$

$$u^3 = \frac{dx^3}{d\tau} = \frac{dz}{dt\sqrt{1-\beta^2}} = \frac{u_z}{\sqrt{1-\beta^2}} \tag{4.57(d)}$$

and are written as

$$u^\mu = \left(\frac{c}{\sqrt{1-\beta^2}}, \frac{\mathbf{u}}{\sqrt{1-\beta^2}}\right) \tag{4.58}$$

The covariant components of u_μ are

$$u_\mu = (u_0, -u_k)$$

$$= \left(\frac{c}{\sqrt{1-\beta^2}}, \frac{-\mathbf{u}}{\sqrt{1-\beta^2}}\right) \tag{4.59}$$

One can easily show that

$$\sum_{\mu=0}^{3} u^\mu u_\mu = c^2 \tag{4.60}$$

Differentiating Eq. (4.60) w.r.t. τ, we have

$$\sum_\mu \frac{du^\mu}{d\tau} u_\mu + \sum_\mu u^\mu \frac{du_\mu}{d\tau} = \sum_\mu \omega^\mu u_\mu + \sum_\mu u^\mu \omega_\mu = 0 \tag{4.61}$$

However, in any pair of dummy indices, we can interchange the places of the superscript and subscript; hence

$$\sum_\mu \omega^\mu u_\mu = \sum_\mu \omega_\mu u^\mu \tag{4.62}$$

Therefore we have

$$\sum_\mu \omega^\mu u_\mu = 0 \tag{4.63}$$

which implies that four-velocity and four-acceleration are orthogonal to each other.

4.5 RELATIVISTIC EQUATION OF MOTION. MINKOWSKI FORCE

In relativistic mechanics, the force acting on a particle is defined as

$$\frac{d\mathbf{p}}{dt} = \mathbf{f} \tag{4.64}$$

where $\mathbf{p} = m\mathbf{u}/(1 - u^2/c^2)^{1/2}$ is the momentum of the particle; m is the rest mass of the particle; and $\mathbf{u}$, the particle velocity. We are using the symbol m instead of m_0, as there is only one kind of mass to be used.

Assuming that the velocity of the particle changes only in direction, i.e., the force is directed perpendicular to the velocity, then we get

$$\frac{d}{dt}\left(\frac{m\mathbf{u}}{\sqrt{1-u^2/c^2}}\right) = \frac{m}{\sqrt{1-u^2/c^2}}\frac{d\mathbf{u}}{dt} \tag{4.65}$$

When the velocity changes only in magnitude, i.e., the force is parallel to the velocity, then

$$\frac{d}{dt}\left(\frac{m\mathbf{u}}{\sqrt{1-u^2/c^2}}\right) = \frac{m}{(1-u^2/c^2)^{3/2}}\frac{d\mathbf{u}}{dt} \tag{4.66}$$

If we formulate the equation of motion in a manner analogous to that in Newtonian mechanics as

mass × Acceleration = Force

then we infer that the particle is endowed with a longitudinal mass $m/(1 - u^2/c^2)^{3/2}$ in the direction of motion and a transverse mass $m(1 - u^2/c^2)^{-1/2}$ in a direction at right angles to its motion. However, the necessity of postulating two masses is circumvented very easily if the equation of motion is stated in the form actually proposed by Newton, i.e.,

Time rate of change of linear momentum = Force

Equation (4.64) is called the fundamental equation of relativistic dynamics. It has been verified experimentally from the study of the bending of the charged particles in electrostatic and magnetostatic fields together, for determining the mass of charged particles.

The simplest way in which Eq. (4.64) can be generalized in a Lorentz covariant manner, i.e., which holds in every Lorentz frame, is by differentiating the momentum four-vector p^μ with respect to τ and obtain another four-vector K^μ called Minkowski force as

$$\frac{d}{d\tau}(mu^\mu) = K^\mu \tag{4.67}$$

where τ is the proper time, m is the invariant mass of the particle, and u^μ is the four-velocity of the particle. We will determine the four-vector and see that the spatial components of Eq. (4.67) for $v << c$ reduce to Eq. (4.64). Rewriting Eq. (4.67) in its component form in view of Eqs (2.59) and (4.59), we get

$$\frac{1}{\sqrt{1-v^2/c^2}}\frac{d}{dt}\left(\frac{mc}{\sqrt{1-v^2/c^2}}\right) = K^0 \tag{4.68}$$

$$\frac{1}{\sqrt{1-v^2/c^2}}\frac{d}{dt}\left[\frac{mv_i}{\sqrt{1-v^2/c^2}}\right] = K_i \; (i = 1, 2, 3) \tag{4.69}$$

or,

$$\frac{d}{dt}\frac{mc}{\sqrt{1-v^2/c^2}} = K^0\sqrt{1-v^2/c^2} \tag{4.70}$$

$$\frac{d}{dt}\left[\frac{mv_i}{\sqrt{1-v^2/c^2}}\right] = K_i\sqrt{1-v^2/c^2} \quad (i = 1, 2, 3) \tag{4.71}$$

Recalling the spatial components of conventional force

$$f_i = K_i\sqrt{1-v^2/c^2} \tag{4.72}$$

Eq. (4.71) may be written as

$$\frac{d}{dt}\left(\frac{mv_i}{\sqrt{1-v^2/c^2}}\right) = f_i \quad (i = 1, 2, 3) \tag{4.73}$$

which agrees with the Newtonian equation, Eq. (4.64). Let us determine the time component K^0 of Minkowski force. Multiplying both sides of Eq. (4.67) by u_μ we get

$$K^\mu u_\mu = m\frac{du^\mu}{d\tau}u_\mu = m\omega^\mu u_\mu \tag{4.74}$$

m being a scalar could be taken outside the derivative sign. Summing it over μ, we get in view of Eq. (4.63)

$$\sum_\mu K^\mu u_\mu = m\sum_\mu \omega^\mu u_\mu = 0 \tag{4.75}$$

Substituting the components of u_μ and K_i from Eqs (4.59) and (4.72), respectively, we get

$$K^0\frac{c}{\sqrt{1-v^2/c^2}} - \sum_{i=1}^{3}\frac{f_i}{\sqrt{1-v^2/c^2}}\frac{v_i}{\sqrt{1-v^2/c^2}} = 0$$

which yields

$$K^0 = \frac{1}{c\sqrt{1-v^2/c^2}}\sum_{i=1}^{3} f_i v_i = \frac{\mathbf{f}.\mathbf{v}}{c\sqrt{1-v^2/c^2}} \tag{4.76}$$

Thus Minkowski force K^μ is

$$K^\mu = \left(\frac{\mathbf{f}.\mathbf{v}}{c\sqrt{1-v^2/c^2}}, \frac{\mathbf{f}}{\sqrt{1-v^2/c^2}}\right) \tag{4.77}$$

Equation (4.76) can be written alternatively in terms of the rate of change of energy of the particle as

$$K^0 = \frac{1}{c\sqrt{1-v^2/c^2}}\frac{dE}{dt} \tag{4.78}$$

where E is the energy of the particle.

Substituting the value of K^0 from Eq. (4.78) in Eq. (4.70) we get

$$\frac{d}{dt}\frac{mc}{\sqrt{1-v^2/c^2}}=\frac{1}{c}\frac{d\mathcal{E}}{dt} \tag{4.79}$$

Thus the relativistic energy of the particle is

$$\mathcal{E}=\frac{mc^2}{\sqrt{1-v^2c^2}}+C_1 \tag{4.80}$$

where C_1 is a constant of integration. The kinetic energy of the particle is zero, when its velocity is zero; hence C_1 is equal to $-m_0c^2$. Therefore,

$$\mathcal{E}=\frac{mc^2}{\sqrt{1-v^2/c^2}}-m_0c^2 \tag{4.81}$$

4.5.1 Transformation of Force

Let us develop the transformations for force **f** and **f′** between two inertial frames S and S', respectively, through the definitions

$$\mathbf{f}=\frac{d\mathbf{p}}{dt} \tag{4.82}$$

$$\mathbf{f}'=\frac{d\mathbf{p}'}{dt'} \tag{4.83}$$

Thus,

$$f'_x=\frac{dp'_x}{dt'}=\frac{dp'_x/dt}{dt'/dt}$$

$$=\frac{\gamma\left(\frac{dp_x}{dt}-\frac{v}{c^2}\frac{d\mathcal{E}}{dt}\right)}{\gamma\left(1-\frac{v}{c^2}\frac{dx}{dt}\right)}=\frac{f_x-\frac{v}{c^2}\frac{d\mathcal{E}}{dt}}{1-\frac{vu_x}{c^2}}$$

where we made use of the following transformations

$$p'_x=\gamma(p_x-v\mathcal{E}/c^2)$$
$$t'=\gamma(t-vx/c^2)$$

Next we show that

$$\frac{d\mathcal{E}}{dt}=\mathbf{f}\cdot\mathbf{u} \tag{4.85}$$

We have

$$\mathcal{E}^2=c^2p^2+m_0^2c^4=c^2(\mathbf{p}\cdot\mathbf{p})+m_0^2c^4 \tag{4.86}$$

Thus, $$\mathcal{E}\frac{d\mathcal{E}}{dt} = c^2\mathbf{p}\cdot\left(\frac{d\mathbf{p}}{dt}\right) \tag{4.87}$$

But $$\mathcal{E} = mc^2$$

Therefore, $$\frac{d\mathcal{E}}{dt} = \frac{c^2}{mc^2}\,\mathbf{p}\cdot\mathbf{f} = \mathbf{f}\cdot\mathbf{u} \tag{4.88}$$

Substituting this in Eq. (4.84), we get the transformation of the force component parallel to the direction of relative motion of two inertial frames as

$$f'_x = \frac{f_x - \dfrac{v}{c^2}\,\mathbf{f}\cdot\mathbf{v}}{1 - \dfrac{vu_x}{c^2}} \tag{4.89}$$

Similarly,

$$f'_y = \frac{dp'_y}{dt'} = \frac{dp'_y}{dt}\bigg/\frac{dt'}{dt}$$

$$= \frac{dp_y/dt}{\gamma\left(1 - \dfrac{vu_x}{c^2}\right)} = \frac{f_y}{\gamma\left(1 - \dfrac{vu_x}{c^2}\right)} \tag{4.90}$$

Analogously we obtain the transformation for f'_z as

$$f'_z = \frac{f_z}{\gamma\left(1 - \dfrac{vu_x}{c^2}\right)} \tag{4.91}$$

The inverse transformations for components of force can be written as

$$f_x = \frac{f'_x + \dfrac{v}{c^2}(\mathbf{f}'\cdot\mathbf{v}')}{1 + \dfrac{vu'_x}{c^2}} \tag{4.92}$$

$$f_y = \frac{f'_y}{\gamma\left(1 + \dfrac{vu'_x}{c^2}\right)} \tag{4.93}$$

$$f_z = \frac{f'_z}{\gamma\left(1 + \dfrac{vu_x}{c^2}\right)} \tag{4.94}$$

Equation (4.89) shows that the measure of force in the frame S' involves the power developed by the force in a plane at right angles to the line of motion in the other frame. It results from the intermingling of space and time in the relativistic description of the event and has no classical analogue. Galileo considered force as the time rate of change of momentum and Huygens as the space rate of change of energy. In the relativistic treatment, both the ideas are unified and appear as partial aspects of a greater concept.

4.6 ENERGY–MASS EQUIVALENCE

Let us determine the relationship between mass and energy. Assuming the relativistic form of Newton's law, we have

$$\mathbf{f} = \frac{d}{dt}(m\mathbf{u}) \tag{4.95}$$

Consider a particle of rest mass m initially at rest but accelerated under a force $\mathbf{f}$ applied in the x-direction. The work done by the force in displacing the particle through distance x_f is

$$T = \int_0^{x_f} f dx = \int_0^{t_f} f \frac{dx}{dt} dt = \int_0^{t_f} fu dt$$

$$= \int_0^{t_f} u \frac{dp}{dt} dt = \int udp$$

$$= up \Big|_0^{u_f} - \int_0^{u_f} pdu$$

Substituting $p = \dfrac{mu}{\sqrt{1 - u^2/c^2}}$ and $udu = 1/2\, d(u^2)$

We get

$$T = \frac{mu^2}{\sqrt{1 - u^2/c^2}} \Bigg|_0^{u_f} - m/2 \int_0^{u_f} \frac{d(u^2)}{\sqrt{1 - u^2/c^2}}$$

$$= mc^2 \left[\frac{u^2/c^2}{\sqrt{1 - u^2/c^2}} + \sqrt{1 - u^2/c^2} \right]_0^{u_f}$$

$$= mc^2 \left[\frac{1}{\sqrt{1 - u^2/c^2}} \right]_0^{u_f}$$

Evaluating at the limits and writing u_f as u, we have

$$T = \frac{mc^2}{\sqrt{1 - u^2/c^2}} - mc^2$$

$$= m_r c^2 - mc^2 = (m_r - m)c^2 \tag{4.96}$$

The kinetic energy is interpreted as the difference in the *relativistic mass* from the rest mass multiplied by c^2. Moreover, the rest mass m is associated with the energy mc^2, called the rest mass energy. The total energy is

$$E = T + mc^2 = m_r c^2 = \frac{mc^2}{\sqrt{1 - u^2/c^2}} \tag{4.97}$$

Equation (4.97) is the law of equivalence of mass and energy or the law of inertia of energy. This constitutes an important basis of the process of conversion of rest energy into other forms of energy, such as in a nuclear reactor or plasma. Thus the two broad principles of classical physics—the law of conservation of energy and the law of conservation of mass—are fused together into this single comprehensive law of conservation of total relativistic energy. This law is an article of faith with the physicists.

4.6.1 Mass Defect: Binding Energy

From Eq. (4.96), we have

$$T = E - E_0 = c^2(m_r - m)$$

$$\Delta\mathcal{E} = \Delta mc^2 \tag{4.98}$$

where $\mathbf{\Delta m}$ is the change in mass corresponding to the change in energy $\mathbf{\Delta\mathcal{E}}$. According to Einstein, this is the most important result of the special theory of relativity. Assuming the conservation of energy in nuclear reactions, the difference between the mass of the nucleus from the sum of the individual nucleons making it must be negative. Calling it ΔM, we have

$$\Delta M = M - \sum_n M_n \tag{4.99}$$

ΔM is called the mass defect and ΔMc^2 is the binding energy of the nucleons which is responsible for holding the nucleus together.

Example 4.3 A nucleus will disintegrate into parts only if the mass defect relative to these parts is positive.

Solution. Let us consider a nucleus of mass M composed of two parts of mass M_1 and M_2 in its inertial frame. Imagine that it disintegrates spontaneously into these parts with velocities $\mathbf{v}_1$ and $\mathbf{v}_2$. According to the relativistic law of energy conservation, we have

$$Mc^2 = \frac{M_1c^2}{\sqrt{1 - v_1^2/c}} + \frac{M_2c^2}{\sqrt{1 - v_2^2/c^2}}$$

This is possible only if

$$M > M_1 + M_2$$

i.e., if

$$\Delta M = M - (M_1 + M_2) > 0$$

Thus the spontaneous disintegration is possible only if the mass defect is positive.

When the total mass of the reactants is greater than that of the products, the energy balance or Q-value is positive, or the reaction is said to be '*exothermic*' or '*exoergic*'. When Q is negative, it is '*endothermic*' or '*endoergic*'.

Example 4.4 One of the earliest quantitative proofs of the validity of the Einstein's mass–energy relationship was provided by Cockroft and Walton[6], the first case of an artificial nuclear disintegration. They bombarded $_3\text{Li}^7$ with $_1\text{H}^1$, accelerated to energies of 0.1–0.7 MeV. The cloud-chamber pictures showed that two α particles leave the point of disintegration and proceed with equal energies in opposite directions. The reaction may be represented as

$$_3\text{Li}^7 + {_1\text{H}^1} \rightarrow [_4\text{Be}^8] \rightarrow {_2\text{He}^4} + {_2\text{He}^4} + Q$$

where Q is the energy balance.

The masses are

$$M\left({}_{3}\mathrm{Li}^{7}\right) = 7.01818 \text{ a.m.u.}$$
$$M\left({}_{1}\mathrm{H}^{1}\right) = 1.008142 \text{ a.m.u.}$$
$$M\left({}_{2}\mathrm{He}^{4}\right) = 4.00386 \text{ a.m.u.}$$

Find the Q-value of the reaction.

Solution. The mass defect

$$\Delta M = 0.01860 \text{ a.m.u.}$$

and

$$Q = \Delta M c^2 = 17.32 \text{ MeV}$$

The experimental value of Q, obtained from the energies of the incident protons and outgoing alpha particles, is 17.33 MeV. This release of energy was at the expense of its mass.

Example 4.5 A body of mass M disintegrates while at rest into two parts of masses M_1 and M_2. Determine the energies $\mathcal{E}_1$ and $\mathcal{E}_2$ of the products.

Solution. Laws of energy and momentum conservation give

$$Mc^2 = \mathcal{E}_1 + \mathcal{E}_2 \qquad \text{(i)}$$

and

$$\mathbf{p}_1 + \mathbf{p}_2 = 0$$

or

$$c^2 p_1^2 = c^2 p_2^2$$

But

$$\mathcal{E}^2 = c^2 p^2 + M^2 c^4$$

Hence

$$c^2 p_1^2 = \mathcal{E}_1^2 - M_1^2 c^4$$

$$c^2 p_2^2 = \mathcal{E}_2^2 - M_2^2 c^4$$

Therefore

$$\mathcal{E}_1^2 - M_1^2 c^4 = \mathcal{E}_2^2 - M_2^2 - c^4$$

or

$$\mathcal{E}_1^2 - \mathcal{E}_2^2 = M_1^2 c^4 - M_2^2 c^4 \qquad \text{(ii)}$$

Dividing Eq. (ii) by Eq. (i), we get

$$\mathcal{E}_1 - \mathcal{E}_2 = \frac{M_1^2 c^4 - M_2^2 c^4}{Mc^2}$$

$$= \frac{M_1^2 c^2 - M_2^2 c^2}{M} \qquad \text{(iii)}$$

Adding Eqs (i) and (iii), we get

$$\mathcal{E}_1 = c^2 \frac{M^2 + M_1^2 - M_2^2}{2M}$$

Also,

$$\mathcal{E}_2 = c^2 \frac{M^2 + M_2^2 - M_1^2}{2M}$$

Example 4.6 A $\pi^{\pm}$ meson of mass m_π comes to rest and disintegrates into a $\mu^{\pm}$ meson of mass m_μ and a neutrino of zero rest-mass. Show that the kinetic energy of motion of the $\mu^{\pm}$ meson is

$$T = \frac{\left(m_\pi - m_\mu\right)}{2m_\pi c^2}$$

Solution. The decay is represented by

$$\pi^{\pm} \rightarrow \mu^{\pm} + v$$

Calling m_π, m_μ, m_v the rest masses of $\pi^{\pm}$, $\mu^{\pm}$, and v respectively and $\mathcal{E}_\pi$, $\mathcal{E}_\mu$, $\mathcal{E}_v$ the total energies of these particles; we have from the law of conservation of energy,

$$\mathcal{E}_\mu^2 - \mathcal{E}_v^2 = \left(\mathcal{E}_\mu^2 - \mathcal{E}_v^2\right) c^4$$

$$m_\pi c^2 = \mathcal{E}_\mu + \mathcal{E}_v \tag{i}$$

since the decay is at rest, the momenta of μ and v must be equal and opposite.
Therefore,

$$\mathcal{E}_\mu - \mathcal{E}_v = \frac{m_\mu^2 - m_v^2}{m_\pi} c^2 \tag{ii}$$

From Eqs (i) and (ii), we get

$$\mathcal{E}_\mu = \frac{m_\mu^2 + m_\pi^2 - m_v^2}{2m_\pi} c^2$$

and

$$\mathcal{E}_v = \frac{m_\pi^2 + m_v^2 - m_\mu^2}{2m_\pi} c^2$$

which reduce to the following equations, since $m_v = 0$.

$$\mathcal{E}_\mu = \frac{m_\mu^2 - m_\mu^2}{2m_\pi} c^2$$

and

$$\mathcal{E}_v = \frac{m_\pi^2 - m_\mu^2}{2m_\pi} c^2$$

The kinetic energy of motion of the μ meson is

$$T = \mathcal{E} - m_\mu c^2$$

$$= \frac{m_\mu^2 + m_\pi^2}{2m_\pi} c^2 - m_\mu c^2$$

$$= \frac{(m_\pi - m_\mu)^2}{2m_\pi} c^2$$

Example 4.7 An excited atom of total mass M_0 at rest with respect to the inertial system gets de-excited into a state with an energy lower by $\Delta\mathcal{E}$. The emitted photon causes the emitting atom to undergo recoil and thus is of lower frequency than $\Delta\mathcal{E}/h$. Compute this frequency.

Solution. Let M_0 be the rest mass of the atom before emission, M'_0 be the rest mass of the atom after emission, and $M' = \frac{M'_0}{\sqrt{1 - v^2/c^2}}$, where v is the recoil velocity.

Let ν be the frequency of the emitted photon. From the laws of conservation of energy and momentum, we get

$$M_0 c^2 = M'c^2 + h\nu \tag{i}$$

$$0 = M'v - \frac{h\nu}{c} \tag{ii}$$

$$M' = \frac{M'_0}{\sqrt{1 - v^2/c^2}} \tag{iii}$$

From Eq. (iii), we get $\quad M'^2 = M_0'^2/(1 - v^2/c^2)$

or $\quad M'^2 v^2 = \left(M'^2 - M_0'^2\right) c^2$

Also from Eq. (ii), $\quad M'^2 v^2 = \left(\frac{h\nu}{c}\right)^2$

Eliminating v between Eqs (ii) and (iii), we have

$$(M'^2 - M_0'^2)c^2 = \left(\frac{h\nu}{c}\right)^2$$

or

$$M'^2 c = \left(\frac{h\nu}{c}\right)^2 + M_0'^2 c^2 \tag{iv}$$

Eliminating M' from Eqs (i) and (iv), we get

$$\left(M_0 c - \frac{h\nu}{c}\right)^2 = M'^2 c^2 = \left(\frac{h\nu}{c}\right)^2 + M_0'^2 c^2$$

Therefore,

$$\left(M_0 c - \frac{h\nu}{c}\right)^2 - \left(\frac{h\nu}{c}\right)^2 = M'_0 c^2$$

or

$$h\nu = \frac{M_0^2 c^2 - M'^2_0 c^2}{2M_0}$$

Now

$$\Delta\mathcal{E} = (M_0 - M'_0)c^2$$

Hence,

$$h\nu = (M_0 - M'_0)\, c^2 \left[\frac{M_0 + M'_0}{2M_0}\right]$$

$$= \Delta\mathcal{E}\left[1 - \frac{M_0 - M'_0}{2M_0}\right]$$

$$= \Delta\mathcal{E}\left[1 - \frac{\Delta\mathcal{E}}{2M_0 c^2}\right]$$

Example 4.8 The gravitational red shift[6]: One of the dramatic illustrations of the use of Mössbauer effect is the measurement of the gravitational red shift, i.e., the change in the 14.4-KeV photon as it travels between regions of different potential. Compute the resultant fractional line-width in 14.4 KeV photon of Fe^{57} when it is allowed to drop through a distance of 74 ft. ($\Delta\mathcal{E}$ for $Fe^{57} = 4.6 \times 10^{-9}$ eV)[7]

Solution. A photon of energy $\mathcal{E}$ has an effective mass $m = \mathcal{E}/c^2$ moving in a gravitational field g. When it drops a distance d, it will gain energy

$$mgd = \frac{\mathcal{E}gd}{c^2} = \Delta\mathcal{E}$$

which gives

$$\frac{\Delta\mathcal{E}}{\mathcal{E}} = \frac{gd}{c^2} = \frac{9.81 \times d}{(3 \times 10^8)^2} = 1.1 \times 10^{-16}\, d$$

$$= 2.48 \times 10^{-15}$$

where d is expressed in metres.

Now for Fe^{57}

$$\Delta\mathcal{E} = 4.6 \times 10^{-9} \text{ eV}$$

$$\mathcal{E} = 14.4 \text{ KeV}$$

Therefore

$$\frac{\Delta\mathcal{E}}{\mathcal{E}} = 3.2 \times 10^{-13}$$

The existence of the effect although very negligible, being only 1 part in 100, has been verified to within 5 per cent.

Example 4.9 A photon of energy $\mathcal{E}_0$ may be described as particle of zero rest mass, with momentum $\mathcal{E}_0/c$ and wavelength $hc/\mathcal{E}_0$. If the photon collides with an electron of mass m_0 at rest, it will be scattered off at some angle ϕ with a new energy $h\nu$. Show that the change in wavelength is related to the scattering angle by the formula

$$\lambda - \lambda_0 = 2\lambda_0 \sin^2\phi/2$$

where $\lambda_c = h/(m_0 c)$ is the Compton wavelength. Also show that the kinetic energy of the recoil motion of the electron is

$$T = h\nu_0 \frac{2(\lambda_c/\lambda_0)\sin^2\phi/2}{1 + 2(\lambda_c/\lambda_0)\sin^2\phi/2}$$

Solution. A photon of energy $h\nu_0$ is incident on an electron, which after scattering gives rise to a scattered photon with energy $h\nu$, whereas the electron which scatters the photon suffers a momentum recoil.

Compton treated this problem of the scattering of a photon by an electron as an elastic collision between them and in this collision energy and momentum are conserved. From the laws of conservation of energy and momentum, Figure 4.2, we get

$$h\nu_0 = h\nu + m_0 c^2\left(\frac{1}{\sqrt{1 - v^2/c^2}} - 1\right) \quad \text{(i)}$$

where **v** is the recoil velocity of the electron and the relativistic energy of the electron is used.

Also,

$$\frac{h\nu_0}{c} = \frac{h\nu}{c}\cos\phi + \frac{m_0 v}{\sqrt{1 - v^2/c^2}}\cos\theta \quad \text{(ii)}$$

$$0 = \frac{h\nu}{c}\sin\phi - \frac{m_0 v}{\sqrt{1 - v^2/c^2}}\sin\theta \quad \text{(iii)}$$

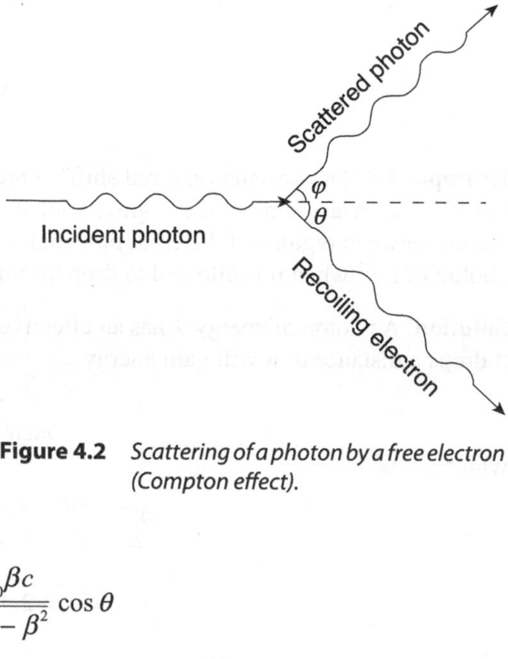

Figure 4.2 *Scattering of a photon by a free electron (Compton effect).*

Introducing $\lambda_0 = c/\nu_0$; $\lambda = c/\nu$ and $\beta = v/c$ of the incident and scattered radiations, respectively, Eqs (ii) and (iii) become

$$\frac{h}{\lambda_0} - \frac{h}{\lambda}\cos\phi = \frac{m_0\beta c}{\sqrt{1 - \beta^2}}\cos\theta$$

$$\frac{h}{\lambda}\sin\phi = \frac{m_0\beta c}{\sqrt{1 - \beta^2}}\sin\theta$$

Eliminating θ from the above equations, by squaring and adding them, we get

$$\frac{h^2}{\lambda_0^2} + \frac{h^2}{\lambda^2} - \frac{2h^2\cos\phi}{\lambda_0\lambda} = \frac{m_0^2\beta^2c^2}{1 - \beta^2} = \frac{m_0^2 c^2}{1 - \beta^2} - m_0^2 c^2 \quad \text{(iv)}$$

Rewriting Eq. (i) in terms of λ_0 and λ, we get

$$\frac{h}{\lambda_0} - \frac{h}{\lambda} + m_0 c = \frac{m_0 c}{\sqrt{1-\beta^2}}$$

Squaring it, we get

$$\frac{h^2}{\lambda_0^2} + \frac{h^2}{\lambda^2} - \frac{2h^2}{\lambda_0 \lambda} + 2m_0 ch\left(\frac{1}{\lambda_0} - \frac{1}{\lambda}\right) + m_0^2 c^2 = \frac{m_0^2 c^2}{1-\beta^2} \qquad \text{(v)}$$

Subtracting Eq. (iv) from Eq. (v), we have

$$\frac{2h^2}{\lambda_0 \lambda}(\cos\phi - 1) + 2m_0 hc\left[\frac{1}{\lambda_0} - \frac{1}{\lambda}\right] = 0$$

or

$$\Delta\lambda = \lambda - \lambda_0 = \frac{h}{m_0 c}(1 - \cos\phi) = 2\,\lambda_c \sin^2 \phi/2$$

Inserting

$$\lambda_c = h/m_0 c$$

$$h = 6.624 \times 10^{-27} \text{ erg-s}$$

$$m_0 = 0.9107 \times 10^{-27} \text{ g}$$

$$c = 2.998 \times 10^{10} \text{ cm/s}$$

we get the change in wavelength

$$\Delta\lambda = 0.0242(1 - \cos\phi)\,\text{Å}$$

As is obvious, the shift in wavelength is independent of the incident wavelength. The quantity $\lambda_c = h/(m_0 c)$ is referred to as the Compton wavelength of the electron.
Change in wavelength $\Delta\lambda$ = expressed in frequency, becomes

$$\Delta\lambda = 2\lambda_c \sin^2 \phi/2$$

or

$$\frac{1}{\nu} - \frac{1}{\nu_0} = \frac{2\lambda_c}{c}\sin^2 \phi/2$$

$$\nu = \frac{\nu_0}{1 + \dfrac{2\lambda_c \nu_0}{c}\sin^2 \phi/2}$$

The kinetic energy of the recoil electron

$$T = h\nu_0 - h\nu$$

$$= h\nu_0\left(1 - \frac{1}{1 + 2\dfrac{\lambda_c \nu_0}{c}\sin^2 \phi/2}\right)$$

$$= h\nu_0 \frac{2\,\frac{\lambda_c \nu_0}{c}\sin^2\phi/2}{1 + 2\,\frac{\lambda_c \nu_0}{c}\sin^2\phi/2}$$

$$= h\nu_0 \frac{2(\lambda_c/\lambda_0)\sin^2\phi/2}{1 + 2(\lambda_c/\lambda_0)\sin^2\phi/2}$$

Example 4.10 Show that it is impossible for a photon to transfer all its energy to a free electron.

Solution. Suppose we have a photon of energy $\mathcal{E}$ and momentum $\mathcal{E}/c$, which transfers all its energy to an electron of mass m and velocity $\mathbf{v}$. Then according to the laws of conservation of energy and momentum, we have

$$\mathcal{E} = \frac{m_0 c^2}{\sqrt{1 - v^2/c^2}} - m_0 c^2$$

$$\Rightarrow \frac{\mathcal{E}}{c} = \frac{m_0 c\,[1 - \sqrt{1 - v^2/c^2}\,]}{\sqrt{1 - v^2/c^2}} \qquad \text{(i)}$$

and

$$\frac{\mathcal{E}}{c} = \frac{m_0 v}{\sqrt{1 - v^2/c^2}} \qquad \text{(ii)}$$

Equations (i) and (ii) lead to

$$\frac{v}{c} = 1 - \sqrt{1 - v^2/c^2}$$

$$\Rightarrow (\sqrt{1 - v^2/c^2})^2 = (1 - v/c)^2$$

or

$$1 - v^2/c^2 = 1 + v^2/c^2 - 2v/c$$

$$\Rightarrow v^2/c^2 = v/c$$

$$\Rightarrow v = c$$

which is an absurd result, since no material particle can travel with a velocity equal to that of light. Thus the assumed process is inadmissible.

Aliter. The same result can be proved through the use of four-vectors. Let k_μ and p_μ be the energy–momentum four-vectors of the photon and electron respectively before the process of absorption, and k'_μ and p'_μ the respective four-vectors after the process. Obviously $k'_\mu = 0$, as there is no photon left after it has been absorbed.

According to the law of conservation of energy–momentum four-vectors, we have

$$k_\mu + p_\mu = p'_\mu \qquad \text{(i)}$$

and the corresponding invariants are

$$k_\mu k^\mu = k^2 = k_0^2 - \mathbf{k}^2 = 0$$

$$p_\mu p^\mu = p^2 = m_0^2 c^2 \qquad \text{(ii)}$$
$$p'_\mu p'^\mu = p'^2 = m_0^2 c^2$$

Writing the four-vectors in the component form, we have

$$k_\mu = \left(\frac{h\nu}{c}, 0, 0, \frac{h\nu}{c}\right)$$

$$p_\mu = \left(\frac{\mathcal{E}_1}{c}, 0, 0, p_1\right) \qquad \text{(iii)}$$

$$p'_\mu = \left(\frac{\mathcal{E}_2}{c}, 0, 0, p_2\right)$$

where we assumed that the energy and momentum of the electron are $\mathcal{E}_1$ and p_1 before the process and $\mathcal{E}_2$ and p_2 after the absorption and that the photon travels along the z-axis. From Eq. (i), we get

$$k_\mu = p'_\mu - p_\mu \qquad \text{(iv)}$$

and since $k_\mu k^\mu = 0$, we have

$$(p'_\mu - p_\mu)(p'^\mu - p^\mu) = 0 \qquad \text{(iv)}$$

or

$$p'^2 + p^2 - 2p_\mu p'^\mu = 2m_0{}^2c^2 - 2(p'_0 p^0 - \mathbf{p}' \cdot \mathbf{p})$$

$$= 2\, m_0 c^2 - 2\left(\frac{\mathcal{E}_1 \mathcal{E}_2}{c^2} - p_1 p_2\right) = 0$$

or

$$\mathcal{E}_1 \mathcal{E}_2 - p_1 p_2 c^2 = m_0^2\, c^4 \qquad \text{(v)}$$

This result can be true only if $\mathcal{E}_1 = \mathcal{E}_2$ and $p_1 = p_2$. This shows that $p_\mu = p'_\mu$ but it contradicts Eq. (i). Thus the assumed process of complete absorption of a photon by an electron is not tenable.

Example 4.11 A π^0 meson of rest energy M_0c^2 moving with velocity v disintegrates into two γ-rays. Calculate the energy distribution of γ-rays from π^0 mesons.

Solution. Let M be the mass of the moving meson π^0; ν_1 and ν_2, the frequencies of the two decay photons, which respectively make angles θ_1 and θ_2 with the initial direction of motion of π^0, Figure 4.3. From the laws of conservation of energy and linear momentum, one gets

$$Mc^2 = h(\nu_1 + \nu_2) \qquad \text{(i)}$$

$$Mv = \frac{h}{c}(\nu_1 \cos\theta_1 + \nu_2 \cos\theta_2) \qquad \text{(ii)}$$

$$0 = \frac{h}{c}(\nu_1 \sin\theta_1 - \nu_2 \sin\theta_2) \qquad \text{(iii)}$$

where

$$M = \frac{M_0}{\sqrt{1 - v^2/c^2}} \qquad \text{(iv)}$$

From Eq. (iv) we have

$$M^2c^2 - M^2 v^2 = M_0^2c^2$$

or

$$M_0^2 c^4 = M^2c^4 - M^2c^2v^2$$

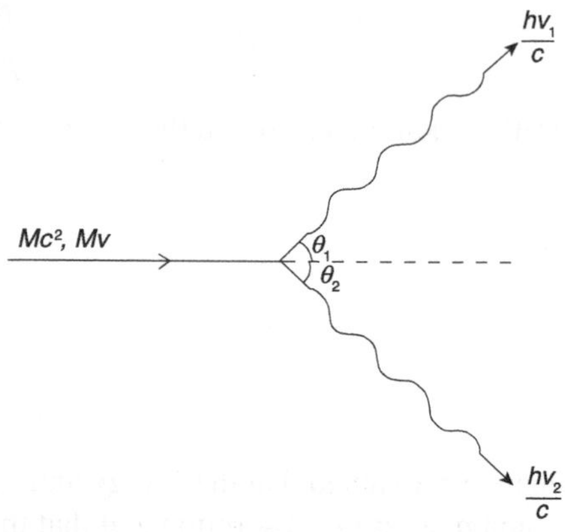

Figure 4.3 *Decay of a π^0 meson into two γ-rays.*

Substituting from Eqs (i) and (ii), we have

$$M_0^2 c^4 = h^2(\nu_1^2 + \nu_2^2 + 2\nu_1\nu_2) - h^2(\nu_1^2 \cos^2\theta^1 + \nu_2^2 \cos^2\theta_2 + 2\nu_1\nu_2 \cos\theta_1 \cos\theta_2)$$

$$= h^2[\nu_1^2\sin^2\theta_1 + \nu_2^2\sin^2\theta_2 + 2\nu_1\nu_2 (1 - \cos\theta_1 \cos\theta_2)]$$

From Eq. (iii), we have

$$\nu_1\sin\theta_1 = \nu_2\sin\theta_2$$

Making use of this relation in the above equation, we get the frequencies as

$$\nu_1^2 = \frac{M_0^2c^4 \sin\theta_2}{2h^2 \sin\theta_1 [1 - \cos(\theta_1 + \theta_2)]}$$

$$\nu_2^2 = \frac{M_0^2 c^4 \sin\theta_1}{2h^2 \sin\theta_2 [1 - \cos(\theta_1 + \theta_2)]}$$

Multiplying the above equations and taking the square root, we have

$$\nu_1\nu_2 = \frac{M_0^2 c^4}{4h^2 \sin^2(\theta_1 + \theta_2)/2}$$

or

$$\sin\left(\frac{\theta_1 + \theta_2}{2}\right) = \frac{M_0c^2}{2h}\sqrt{\frac{1}{\nu_1\nu_2}} = \frac{M_0c^2}{2}\frac{1}{\sqrt{\mathcal{E}_1\mathcal{E}_2}}$$

where $h\nu_1 = \mathcal{E}_1$ and $h\nu_2 = \mathcal{E}_2$.

This gives the energy distribution of the decay photons.

Aliter. Let us solve the problem of π^0-meson decay through the use of four-vectors. The disintegration process is shown schematically in the above diagram. Denoting the four-vectors of the decay photons as

$$k^\mu = \left(\frac{h\nu_1}{c}, \frac{h\nu_1}{c}\mathbf{n}_1\right), \qquad q^\mu = \left(\frac{h\nu_2}{c}, \frac{h\nu_2}{c}\mathbf{n}_2\right)$$

where

$$\frac{\mathbf{n}_1\cdot\mathbf{p}}{|\mathbf{p}|} = \cos\theta_1, \qquad \frac{\mathbf{n}_2\cdot\mathbf{p}}{|\mathbf{p}|} = \cos\theta_2$$

$$\mathbf{n}_1\cdot\mathbf{n}_2 = \cos(\theta_1 + \theta_2)$$

According to the law of conservation of energy–momentum, we have

$$p_\mu = k_\mu + q_\mu$$

which gives

$$p_\mu p^\mu = (k_\mu + q_\mu)(k^\mu + q^\mu)$$

$$M_0^2 c^2 = k^\mu k_\mu + q^\mu q_\mu + 2q^\mu k_\mu$$

$$= 2h^2 \left(\frac{\nu_1 \nu_2}{c^2} - \frac{\nu_1 \nu_2}{c^2} \mathbf{n}_1 \cdot \mathbf{n}_2 \right)$$

or

$$M_0^2 c^4 = 2h^2 \nu_1 \nu_2 [1 - \cos(\theta_1 + \theta_2)]$$

which, on simplification, yields

$$\sin^2 \left(\frac{\theta_1 + \theta_2}{2} \right) = \frac{M_0^2 c^4}{4h^2} \frac{1}{\nu_1 \nu_2}$$

or

$$\sin \frac{\theta_1 + \theta_2}{2} = \frac{M_0 c^2}{2} \sqrt{\frac{1}{\mathcal{E}_1 \mathcal{E}_2}}$$

where $\mathcal{E}_1 = h\nu_1$ and $\mathcal{E}_2 = h\nu_2$.

Example 4.12 A particle with kinetic energy T_0 and rest energy $\mathcal{E}_0$ strikes an identical particle at rest and gets scattered at an angle θ. Show that its kinetic energy T after scattering is given by

$$T = \frac{T_0 \cos^2 \theta}{1 + (T_0 \sin^2 \theta)/2\mathcal{E}_0}$$

Solution. Let a particle of mass m_1 and momentum $\mathbf{p}_1$ strike a particle of identical mass at rest and get scattered at an angle θ_3 with momentum $\mathbf{p}_3$ and scattering the target particle at an angle θ_4 carrying the momentum $\mathbf{p}_4$, as shown in Figure 4.4.
We know that

$$T = mc^2 - m_0 c^2 = m_0 c^2 \left(\frac{1}{\sqrt{1 - v^2/c^2}} - 1 \right)$$

$$(T + \mathcal{E}_0)^2 = \mathcal{E}^2 = c^2 p^2 + m_0^2 c^4 = c^2 p^2 + \mathcal{E}_0^2$$

or

$$p^2 = \frac{2T\mathcal{E}_0}{c^2} + \frac{T^2}{c^2}$$

From the laws of conservation of linear momentum and energy, we have

$$p_1 = p_3\cos\theta_3 + p_4\cos\theta_4 \qquad \text{(i)}$$

$$0 = p_3\sin\theta_3 - p_4\sin\theta_4 \qquad \text{(ii)}$$

$$\mathcal{E}_1 + \mathcal{E}_2 = \mathcal{E}_3 + \mathcal{E}_4$$

or

$$\mathcal{E}_0 + T_1 + \mathcal{E}_0 = \mathcal{E}_0 + T_3 + \mathcal{E}_0 + T_4$$

so

$$T_1 = T_3 + T_4 \qquad \text{(iii)}$$

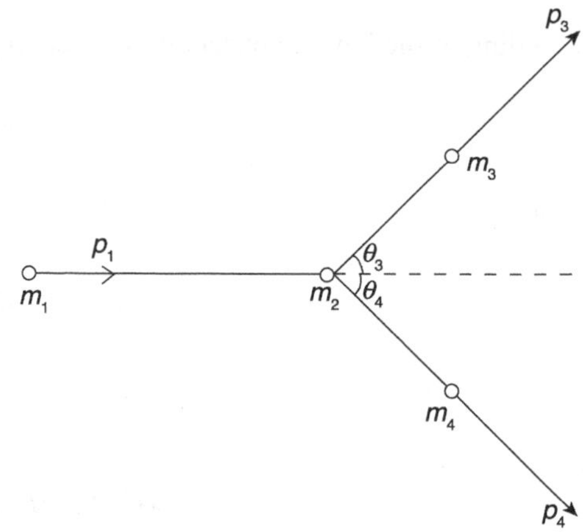

Figure 4.4 *Elastic scattering of particles.*

From Eqs (i) and (ii)

$$p_4^2c^2 = \left(p_1^2 + p_3^2 - 2p_1p_3\cos\theta_3\right)c^2$$

Employing the result

$$p^2 = \frac{2T\mathcal{E}_0}{c^2} + \frac{T^2}{c^2}$$

$$2T_4\mathcal{E}_0 + T_4^2 = 2T_1\mathcal{E}_0 + T_1^2 + 2T_3\mathcal{E}_0 + T_3^2 - 2\sqrt{2T_1\mathcal{E}_0 + T_1^2}\,\sqrt{2T_3\mathcal{E}_3 + T_3^2}\,\cos\theta_3$$

$$2(T_1 - T_3)\,\mathcal{E}_0 + (T_1 - T_3)^2 = 2T_1\mathcal{E}_0 + T_1^2 + 2T_3\mathcal{E}_0 + T_3^2 - 2\sqrt{2T_1\mathcal{E}_0 + T_1^2}\,\sqrt{2T_3\mathcal{E}_3 + T_3^2}\,\cos\theta_3$$

or

$$T_3(T_1 + 2\mathcal{E}_0) = \sqrt{2T_1\mathcal{E}_0 + T_1^2}\,\sqrt{2T_3\mathcal{E}_3 + T_3^2}\,\cos\theta_3$$

$$T_3^2\,(T_1 + 2\mathcal{E}_0)^2 = T_1\,(T_1 + 2\mathcal{E}_0)\,T_3\,(2\mathcal{E}_0 + T_3)\cos^2\theta_3$$

$$T_3 = \frac{2T_1\mathcal{E}_0\cos^2\theta_3}{2\mathcal{E}_0 + T_1\,(1 - \cos^2\theta_3)} = \frac{T_1\cos^2\theta_3}{1 + \dfrac{T_1\sin^2\theta_3}{2\mathcal{E}_0}}$$

4.7 CENTRE OF MOMENTUM TRANSFORMATION

A common problem in nuclear or high energy physics is the production of two or more particles by collision between a projectile (particle 1) with mass m_1, momentum $\mathbf{p}_1 = \mathbf{p}$, and energy $\mathcal{E}_1$, impinging on the target (particle 2) of mass m_2 at rest in the laboratory.

We will relate the incident energy and momentum in the laboratory to the *C-M* variables, by employing the concept of the invariance of the scalar product of two four-vectors. Thus,

$$\mathcal{E}^2/c^2 - p^2 = \mathcal{E}'^2/c^2 - p'^2 \qquad (4.100)$$

where the left hand-side is to be evaluated in the laboratory, the right hand-side in the *C-M* system where $\mathbf{p}' = \mathbf{p}_1' + \mathbf{p}_2' = 0$.

Now

$$\mathcal{E} = \mathcal{E}_1 + m_2c^2$$

and

$$\mathcal{E}' = \mathcal{E}_1' + \mathcal{E}_2'$$

Therefore,

$$\left(\mathcal{E}_1 + m_2c^2\right)^2 - c^2p^2 = \left(\mathcal{E}_1' + \mathcal{E}_2'\right)^2$$

But,

$$\mathcal{E}_1^{\,2} = c^2p^2 + m_1^{\,2}c^4$$

so

$$\mathcal{E}' = \left(m_1^2 + m_2^2 + \frac{2m_2\mathcal{E}_1}{c^2}\right)^{\frac{1}{2}} c^2 \tag{4.101}$$

Defining

$$\Delta Mc^2 = \left(m_1' + m_2'\right)c^2 - \left(m_1 + m_2\right)c^2 \tag{4.102}$$

The criterion for the production just to occur is that there be enough energy available in the *C-M* system that the particles be created with no kinetic energy. This implies that

$$(\mathcal{E}')_{th} = m_1c^2 + m_2c^2 + \Delta Mc^2 \tag{4.103}$$

Substituting this value of $(\mathcal{E}')_{th}$ from Eq. (4.103) in Eq. (4.101), we get

$$\left(m_1 + m_2 + \Delta M\right)c^2 = \left(m_1^2 + m_2^2 + \frac{2m_2\,\mathcal{E}_1}{c^2}\right)^{\frac{1}{2}} c^2 \tag{4.104}$$

Hence the incident kinetic energy of the projectile at threshold is

$$T_{th} = \left(\mathcal{E}_1 - m_1c^2\right)$$

$$= \Delta M\left(1 + \frac{m_1}{m_2} + \frac{\Delta M}{2m_2}\right)c^2 \tag{4.105}$$

Example 4.13 To produce a 'proton–antiproton' pair, an energy of $2M_0c^2 = 1877$ MeV is required in the *C-M* system. What is the minimum energy of (i) a proton for producing a proton pair and (ii) photon needed to produce a π-meson; by striking a proton at rest.

Solution.

(i) Now consider the production of a proton–antiproton pair in proton–proton collisions. Thus the reaction is represented as

$$p + p \rightarrow p + p + p + \overline{p}$$

where p stands for proton and $\bar{p}$ for anti-proton. p and $\bar{p}$ are produced in pairs for conserving the baryon number.
The mass difference is

$$\Delta Mc^2 = 2\, m_p c^2 = 1.877 \text{ BeV}$$

Hence from Eq. (4.105), we get

$$T_{th} = 2\, m_p c^2 (1 + 1 + 1) = 5.631 \text{ BeV}$$

(ii) The reaction is represented by

$$\gamma + p \rightarrow \pi^0 + p$$

The photon has no rest mass. The mass difference is

$$\Delta Mc^2 = m_\pi^0 c^2 = 135.5 \text{ MeV}$$

while the target mass $m_p c^2 = 938.5$ MeV.
Hence,

$$T_{th} = 135.5\left[1 + \frac{135.5}{2(938.5)}\right] = 135.5(1.072) = 144.7 \text{ MeV}$$

Example 4.14 The particle densities (i.e., the number of particles per unit volume) of the colliding beams are n_1 and n_2 respectively. If the difference between the velocities of the particles in the two beams is denoted by $u_{12} = |\mathbf{u}_2 - \mathbf{u}_1|$, then show that

$$\frac{\sigma u_{12}}{1 - \frac{\mathbf{u}_1 \cdot \mathbf{u}_2}{c^2}} = \text{invariant}$$

where σ is the effective collision cross section. Derive an expression for the number of collisions in terms of the effective cross-section σ_0 in a frame of reference in which one of the colliding particles is at rest.

Solution. The number of collisions in a volume dV for time dt is

$$dv = \sigma u_{12}\, n_1 n_2 dVdt \tag{i}$$

which is an invariant quantity. Since $dV\, dt$ is a four-dimensional scalar, we infer that

$$\sigma\, u_{12} n_1 n_2 = \text{invariant} \tag{ii}$$

Further, the number of particles in a given volume element dV given by ndV is an invariant quantity

$$ndV = n_0 dV_0 \tag{iii}$$

Since the transformation is given by

$$dV = dV_0 \sqrt{1 - v^2/c^2}$$

we get the transformation law for the particle density as

$$n = \frac{n_0}{\sqrt{1 - v^2/c^2}} \tag{iv}$$

Rewriting Eq. (ii) in the light of Eq. (iv), we get

$$\frac{\sigma u_{12}}{\sqrt{1 - u_1^2/c^2}\,\sqrt{1 - u_2^2/c^2}} = \text{invariant} \tag{v}$$

The product of the four-velocities of the particles in the two beams is

$$(u_{\mu 1}, u_{\mu 2}) = \frac{1 - (\mathbf{u}_1 \cdot \mathbf{u}_2)/c^2}{\sqrt{1 - u_1^2/c^2}\,\sqrt{1 - u_2^2/c^2}} \tag{vi}$$

and it is an invariant quantity being the product of two four-vectors. Dividing Eq. (v) by Eq (vi), we get

$$\frac{\sigma u_{12}}{1 - \dfrac{\mathbf{u}_1 \cdot \mathbf{u}_2}{c^2}} = \text{invariant} \tag{vii}$$

This was first deduced by Pauli (1933).
Writing Eq. (vii) in terms of effective cross section σ_0, we have

$$\frac{\sigma u_{12}}{1 - \dfrac{\mathbf{u}_1 \cdot \mathbf{u}_2}{c^2}} = \sigma_0 u_{rel} \tag{viii}$$

But relative velocity is given by the following (Example 2.9):

$$u_{rel} = \frac{\sqrt{(\mathbf{u}_1 - \mathbf{u}_2)^2 - (\mathbf{u}_1 \times \mathbf{u}_2)^2/c^2}}{1 - (\mathbf{u}_1 \cdot \mathbf{u}_2)/c^2} \tag{ix}$$

Comparing Eqs (viii) and (ix), we get

$$\frac{\sigma u_{12}}{1 - (\mathbf{u}_1 \cdot \mathbf{u}_2)/c^2} = \sigma_0 \frac{\sqrt{(\mathbf{u}_1 - \mathbf{u}_2)^2 - (\mathbf{u}_1 \times \mathbf{u}_2)^2/c^2}}{1 - (\mathbf{u}_1 \cdot \mathbf{u}_2)/c^2}$$

or

$$\sigma u_{12} = \sigma_0 \sqrt{(\mathbf{u}_1 - \mathbf{u}_2)^2 - (\mathbf{u}_1 \times \mathbf{u}_2)^2/c^2}$$

Substituting it in Eq. (i), we get

$$d\nu = \sigma_0 \sqrt{(\mathbf{u}_1 - \mathbf{u}_2)^2 - (\mathbf{u}_1 \times \mathbf{u}_2)^2/c^2}\; n_1 n_2 dV dt$$

which gives the number of collisions in terms of effective cross section σ_0 in a frame in which one of the colliding particles is at rest.

4.8 KINEMATIC RELATIONS FOR COLLISIONS

Consider an elastic collision between two particles, that is, a collision in which the internal quantum state of neither particle gets changed due to it. Defining $\mathbf{p}_{10}$, $\mathcal{E}_{10}$ as the momentum and energy of the first particle in system S before the collision; $\mathbf{p}_{20}$, $\mathcal{E}_{20}$ pertaining to the second particle and denoting the corresponding quantities in system S' by $\mathcal{E}_1'$, $\mathcal{E}_2'$, $\mathbf{p}_1' = -\mathbf{p}_2'$; S' is obviously the centre of momentum system. Let the x-axis of the S system be along the direction of the total momentum of the particles, $\mathbf{p}_{10} + \mathbf{p}_{20}$.

As usual, we will be working in the 'centre of mass' coordinate system, or better still the 'zero-momentum' reference system. We have

$$\mathcal{E}_{10} + \mathcal{E}_{20} = \frac{\mathcal{E}_1' + \mathcal{E}_2'}{\sqrt{1 - v^2/c^2}} \tag{4.106}$$

$$p_{10x} + p_{20x} = \frac{v(\mathcal{E}_1' + \mathcal{E}_2')/c^2}{\sqrt{1 - v^2/c^2}} = \frac{v}{c^2}\left(\mathcal{E}_{10} + \mathcal{E}_{20}\right) \tag{4.107}$$

$$p_{10y} + p_{20y} = 0 \tag{4.108}$$

$$p_{10z} + p_{20z} = 0 \tag{4.109}$$

Writing vectorially,

$$\mathbf{p}_{10} + \mathbf{p}_{20} = \frac{\mathbf{v}}{c^2}\left(\mathcal{E}_{10} + \mathcal{E}_{20}\right) \tag{4.110}$$

Thus the velocity of the S' system relative to S is

$$\mathbf{v} = \frac{\left(\mathbf{p}_{10} + \mathbf{p}_{20}\right)c^2}{\mathcal{E}_{10} + \mathcal{E}_{20}} \tag{4.111}$$

Let us consider a collision in which the particle with mass m_2 and energy $\mathcal{E}_{20}$ is at rest in S system before the collision. The observer in the S' system will construe that it has velocity $-\mathbf{v}$ along the x'- axis and momentum $-m_2 v/\sqrt{1 - v^2/c^2}$.

In the S' system $\mathbf{p}_{10}' + \mathbf{p}_{20}' = 0$, that is, the momenta are equal in magnitude and opposite in direction. By virtue of the law of conservation of momentum, these will remain equal and opposite but collision could cause these to rotate.

Let ϕ be the angle of rotation of $\mathbf{p}_{10}'$ and $\mathbf{p}_{20}'$ by the collision in the S' system. The x' component of the momentum of the particle with mass m_2 after the collision is $-m_2 v/\sqrt{1 - v^2/c^2}$ and its energy in the S' system is

$$\mathcal{E}_2' = \frac{m_2 c^2}{\sqrt{1 - v^2/c^2}}$$

Its energy in the S system after collision,

$$\mathcal{E}_2 = \frac{\mathcal{E}_{20}' + vp_{10x}'}{\sqrt{1 - v^2/c^2}} = \frac{\dfrac{m_2 c^2}{\sqrt{1 - v^2/c^2}} - \dfrac{m_2 v^2 \cos\phi}{\sqrt{1 - v^2/c^2}}}{\sqrt{1 - v^2/c^2}}$$

$$= \frac{m_2 c^2\left(1 - v^2 \cos\varphi/c^2\right)}{1 - v^2/c^2} \tag{4.112}$$

Eliminating the auxiliary quantity **v** by using Eq. (4.111),

$$\mathbf{v} = \frac{(\mathbf{p}_{10} + \mathbf{p}_{20})c^2}{\mathcal{E}_{10} + \mathcal{E}_{20}}$$

$$= \frac{\mathbf{p}_{10}c^2}{\mathcal{E}_{10} + m_2c^2} \tag{4.113}$$

Since $\mathbf{p}_{20} = 0$ and $\mathcal{E}_2 = m_2c^2$ before collision, therefore

$$v = c\,\frac{\sqrt{\mathcal{E}_{20}^2 - m_1^2 c^4}}{\mathcal{E}_{10} + m_2c^2} \tag{4.114}$$

Now,

$$v^2 = \frac{c^2\left(\mathcal{E}_{20}^2 - m_1^2 c^4\right)}{\left(\mathcal{E}_{10} + m_2c^2\right)^2}$$

or

$$\frac{v^2}{c^2} = \frac{\left(\mathcal{E}_{10}^2 - m_1^2 c^4\right)}{\left(\mathcal{E}_{10} + m_2c^2\right)^2} \tag{4.115}$$

From Eq. (4.112), we get

$$\mathcal{E}_2 = \frac{m_2c^2\left[1 - \dfrac{\mathcal{E}_{10}^2 - m_1^2 c^4}{\left(\mathcal{E}_{10} + m_2c^2\right)^2}\cos\phi\right]}{1 - \dfrac{\mathcal{E}_{10}^2 - m_1^2 c^4}{\left(\mathcal{E}_{10} + m_2c^2\right)^2}}$$

$$= \frac{m_2c^2\left[\left(\mathcal{E}_{10} + m_2c^2\right)^2 - \left(\mathcal{E}_{10}^2 - m_1^2 c^4\right) + \left(\mathcal{E}_{10}^2 - m_1^2 c^4\right) + (1 - \cos\phi)\right]}{\left(\mathcal{E}_{10} + m_2c^2\right)^2 - \left(\mathcal{E}_{10}^2 - m_1^2 c^4\right)}$$

$$= m_2c^2 + \frac{m_2\left(\mathcal{E}_{10}^2 - m_1^2 c^4\right)}{m_1^2 c^2 + m_2^2 c^2 + 2m_2\mathcal{E}_{10}}(1 - \cos\phi) \tag{4.116}$$

The second term represents the energy acquired by the second particle from the first during the collision. Employing the law of conservation of energy, that is, the sum of the energies of the two particles is conserved in the collision, we infer that the energy of the first particle after the collision is given by

$$\mathcal{E}_1 = \mathcal{E}_{10} - \frac{m_2\left(\mathcal{E}_{10}^2 - m_1^2 c^4\right)}{m_1^2 c^2 + m_2^2 c^2 + 2m_2\mathcal{E}_{10}}(1 - \cos\phi) \tag{4.117}$$

Obviously, the maximum transfer of energy occurs when the second term in Eq. (4.116) is the largest, which will be the case for $\varphi = \pi$. Thus,

$$\frac{2m_2\left(\mathcal{E}_{10}^2 - m_1^2 c^4\right)}{\left(m_1^2 c^2 + m_2^2 c^2 + 2m_2 \mathcal{E}_{10}\right) \mathcal{E}_{10}} = \frac{2m_2\left(1 - m_1^2 c^4/\mathcal{E}_{10}^2\right)}{\left[\left(m_1^2 c^2 + m_2^2 c^2/\mathcal{E}_{10}\right)\right] + 2m_2}$$

$$= \frac{\left(1 - m_1^2 c^4/\mathcal{E}_{10}^2\right)}{\left(m_1^2 c^2 + m_2^2 c^2\right)/2m_2\mathcal{E}_{10} + 1} \approx 1 \tag{4.118}$$

Also,

$$\mathcal{E}_{1\min} = \mathcal{E}_{10} - \frac{2m_2\left(\mathcal{E}_{10}^2 - m_1^2 c^4\right)}{m_1^2 c^2 + m_2^2 c^2 + 2m_2\mathcal{E}_{10}}$$

$$= \mathcal{E}_{10} - \frac{\left(\mathcal{E}_{10} - m_1c^2\right)\left(\mathcal{E}_{10} + m_1c^2\right)2m_2}{m_1^2 c^2 + m_2^2 c^2 + 2m_2 \mathcal{E}_{10}}$$

$$= \mathcal{E}_{10} - \mathcal{E}_{10} + m_1c^2 + \frac{\left(\mathcal{E}_{10} - m_1c^2\right)\left(m_1^2 + m_2^2 - 2m_1m_2\right)c^2}{m_1^2 c^2 + m_2^2 c^2 + 2m_2\mathcal{E}_{10}}$$

$$= m_1c^2 + \frac{\left(\mathcal{E}_{10} - m_1c^2\right)\left(m_2 - m_1\right)^2 c^2}{m_1^2 c^2 + m_2^2 c^2 + 2m_2 \mathcal{E}_{10}} \tag{4.119}$$

Let us study the behaviour of Eq. (4.119) under limiting conditions. Rewriting it as

$$\frac{\mathcal{E}_{1\min} - m_1c^2}{\mathcal{E}_{10} - m_1c^2} = \frac{\left(m_2 - m_1\right)^2}{m_1^2 + m_2^2 + 2m_2 \mathcal{E}_{10}/c^2} \tag{4.120}$$

This gives the ratio of the minimum kinetic energy of the impinging particle to the initial kinetic energy.

When $v << c$, i.e. when $\mathcal{E}_{10} \cong m_1c^2 + \frac{1}{2} m_1v^2 + \cdots$, it becomes

$$\frac{\mathcal{E}_{1\min} - m_1c^2}{\mathcal{E}_{10} - m_1c^2} = \frac{\left(m_2 - m_1\right)^2}{\left(m_1 + m_2\right)^2} \tag{4.121}$$

In the limiting case, when $v \sim c$, so that when $\mathcal{E}_{10}$ is very large compared with m_1c^2, Eq. (4.119) reduces to

$$\mathcal{E}_{1\min} = m_1c^2 + \frac{\left(1 - m_1c^2/\mathcal{E}_{10}\right)\left(m_2 - m_1\right)^2 c^2}{\left[\left(m_1^2 c^2 + m_2^2 c^2\right)/\mathcal{E}_{10}\right] + 2m_2}$$

$$\cong \frac{m_1^2 + m_2^2}{2m_2} c^2 \tag{4.122}$$

When $m_2 >> m_1$, the fraction of energy transferred in collision of a light particle with a very heavy target is

$$\frac{2m_2\left(\mathcal{E}_{10}^2 - m_1^2 c^4\right)}{\left(m_1^2 c^2 + m_2^2 c^2 + 2m_2 \mathcal{E}_{10}\right)\mathcal{E}_{10}} = \frac{2m_2(1 - m_1^2 c^4/\mathcal{E}_{10}^2)}{\left[\left(m_1^2 c^2 + m_2^2 c^2\right)/\mathcal{E}_{10}\right] + 2m_2}$$

$$= \frac{(1 - m_1^2 c^4/\mathcal{E}_{10}^2)}{[(m_1^2 c^2 + m_2^2 c^2)/2m_2 \mathcal{E}_{10}] + 1} \approx 1 \tag{4.123}$$

provided $\mathcal{E}_{10} \approx m_2 c^2$.

So the light particle must have energy of the order of the rest energy of the heavy particle. The result is quite unlike the one expected from classical mechanics, which predicts only a negligible transfer of energy from the light particle to the heavy target. When $m_2 << m_1$, which implies that a heavy particle is impinging on a light target, we have

$$\mathcal{E}_{1\min} \approx m_1^2 c^2/m_2 \tag{4.124}$$

Let us derive a relation between the angle of scattering of the particle in S system and handle the problem by four-vector operation. For each of the particles, we have the relation

$$\left(p_{\mu 0} - p_\mu\right) u_\mu = 0 \tag{4.125}$$

where $p_{\mu 0}$, p_μ are the four-momentum of the particle before and after the collision, u_μ is the velocity four-vector, whose space components in system S coincide with the velocity **v**. As the quantity on the left-hand side is a scalar, it will suffice to verify it in any system of reference. In system S', $u_i = 0$, so that $(p_{\mu 0} - p_\mu) = 1/c\ (\mathcal{E}_0 - \mathcal{E})$. However, in system S' the energy of the particle is not changed by collision, that is, $\mathcal{E}_0 = \mathcal{E}$ which proves the relation, Eq. (4.125).

The particle m_2 is initially at rest in the S system and let θ_2 be the angle of scattering. Substituting the values of the components of the vectors $p_{\mu 0}$, p_μ, u_μ in the system S in Eq. (4.125), we get

$$\mathcal{E}_2 - m_2 c^2 = v p_2 \cos\theta_2$$

But from Eq. (4.111), we have

$$v = \frac{p_{10} c^2}{\mathcal{E}_{10} + m_2 c^2} \tag{4.126}$$

Hence we get

$$\cos\theta_2 = \frac{\left(\mathcal{E}_{10} + m_2 c^2\right)\left(\mathcal{E}_2 - m_2 c^2\right)}{p_{10} p_2 c^2} \tag{4.127}$$

Analogously, substituting the values of the four-vector for the impinging particle in Eq. (4.125), we get

$$\mathcal{E}_{10} - \mathcal{E}_1 = v\left(p_{10} - p_1 \cos\theta_1\right) \tag{4.128}$$

which upon substituting $v = p_{10}c^2/(\mathcal{E}_{10} + m_2c^2)$, becomes

$$\cos\theta_1 = \frac{\mathcal{E}_1(\mathcal{E}_{10} + m_2c^2) - \mathcal{E}_{10}\, m_2c^2 - m_1^2\, c^4}{p_{10}p_1c^2} \tag{4.129}$$

When the impinging particle is a photon, then setting $m_1 = 0$; $p_{10} = \mathcal{E}_{10}/c$; $p_1 = \mathcal{E}_1/c$, Eqs (4.127) and (4.129) reduce to

$$\cos\theta_1 = \frac{\mathcal{E}_1\,(\mathcal{E}_{10} + m_2c^2) - \mathcal{E}_{10}m_2c^2}{\mathcal{E}_{10}\mathcal{E}_1} \tag{4.130}$$

$$\cos\theta_2 = \frac{(\mathcal{E}_{10} + m_2c^2)\,(\mathcal{E}_2 - m_2c^2)}{\mathcal{E}_{10}p_2c} \tag{4.131}$$

In Newtonian mechanics, when two equal-mass objects scatter elastically, in the rest frame of one initial particle, the two outgoing particles always travel at right angles to each other. In special relativity, that angle can be much less than a right angle, and in this experiment,[8] it is strikingly less than 90°.

References

1. S. L. Synge, *Relativistic Dynamics, Handbuch der Physik*, Vol. III, Part 1, 198–225, S. Flugge.
2. D. Park, "Relativistic mechanics of a particle", *Am. J. Phys.* (1959) *27*, 311. Axiomatic development to parallel Newtonian mechanics.
3. N. S. Kalitsin, "Relativistic mechanics of a material point of variable mass", *JETP(1955)1*, 565–567.
4. A. O., Barut, "On covariant canonical formulation of relativistic mechanics", *Am. J. Phys.* (1955) *23*, 392.
5. W. Kaufmann, *Ann. der Phys.* (1906) *19*, 487.
6. J. D. Cockroft and E. T. S. Walton, *Proc. Roy. Soc. Lond. A* (1932) *137*, 229. Also K.T. Bainbridge, *Phys. Rev.* (1933) *44*, 123; M. LE. Oliphant, B. B. Kinsey and E. Rutherford, *Proc. Roy. Soc.* (1933) *141*, 722–733.
7. R. V. Pound and G. A. Rebka, *Phys. Rev. Lett.* (1960) *4*(7), 337.
8. Akerlof et al., "Elastic proton-proton scattering at 90°.and structure within the proton", *Phys. Rev.* (1967) *159*, 1138.

Problems

4.1 Show which of the following reactions are exoergic or endoergic and compute the Q-values. These are

$${}_{29}Cu^{65} + {}_1H^1 \rightarrow [{}_{30}Zn^{66}] \rightarrow {}_{30}Zn^{65} + {}_0n^1$$

$${}_{13}Al^{27} + {}_1H^2 \rightarrow [{}_{14}Si^{29}] \rightarrow {}_{12}Mg^{25} + {}_2He^4$$

$${}_{13}Al^{27} + {}_0n^1 \rightarrow [{}_{13}Al^{28}] \rightarrow {}_{12}Mg^{27} + {}_1H^1$$

$${}_1H^2 + \gamma \rightarrow {}_1H^1 + {}_0n^1$$

4.2 Compute the annual loss in mass of sun, if approximately 2 cal of radiated energy are received by each square centimetre of the earth's surface per minute. The distance from earth to sun is about 15×10^7 km.

4.3 Determine the threshold kinetic energies in MeV for the following reactions:

(i) Nucleon pair production in electron–electron collisions,
(ii) π-meson production in π-meson–nucleon collisions.

4.4 A cosmic-ray particle of energy 100 BeV strikes a weight of 1 mg from below. Assuming that the particle transfers all its energy to the weight, show that it will raise the weight by 1.6 mm. (Such a thing cannot happen in practice, because the particle would go clearly through the weight.)

4.5 A particle with initial energy $\mathcal{E}$ and rest energy $\mathcal{E}_0$ hits an identical particle at rest. Assuming that $\mathcal{E} >> \mathcal{E}_0$, show that the maximum energy available in the zero momentum frame is $\sqrt{2\mathcal{E}\mathcal{E}_0}$.

4.6 Show that the components of acceleration in the primed system S' for acceleration $\mathbf{a}(a_x, a_y, a_z)$ in S are

$$a'_x = \frac{a_x}{\gamma^3(1 - vu_x/c^2)^3}$$

$$a'_y = \frac{a_y(1 - vu_x/c^2) + vu_x a_x}{\gamma^2(1 - vu_x/c^2)^3}$$

$$a'_z = \frac{a_z(1 - vu_x c^2) + vu_x a_x c^2}{\gamma^2(1 - vu_x/c^2)^3}$$

4.7 C^{12} nucleus consists of six protons and six neutrons held together by nuclear forces. Find the binding energy of a C^{12} nucleus. Given:

$M_{C^{12}} = 12.000$ a.m.u.; $m_p = 1.007825$ a.m.u.; $m_n = 1.8665$ a.m.u. (92.17 MeV)

4.8 A body is initially at rest. Fifty per cent of its mass is destroyed and given as kinetic energy to the other half. What is the resulting velocity? $(0.866c)$

4.9 Show that the energy relation $\mathcal{E}^2 = c^2p^2 + m_0^2 c^4$, is Lorentz invariant.

4.10 Calculate the fractional change in the mass of hydrogen atom when it is ionized. Given that binding energy of hydrogen atom = 13.58 eV and mH = 1.00797 a.m.u. $(1.44 \times 10^{-6}\%)$

4.11 For the reaction

$$\pi^-(\text{at rest}) + p\ (\text{at rest}) \rightarrow n + \gamma$$

find the kinetic energy T_3 for n and T_4 for γ. Given that

$$m_\pi = 139.6 \text{ MeV};\ m_p = 938.02 \text{ MeV};\ m_n = 939.5 \text{ MeV}$$

[Hint: $T_3 = \left[(m_1 + m_2 - m_3)^2 - m_4^2\right]/2m_1$

$T_4 = \left[(m_1 + m_2 - m_4 - m_3^2)\right]/2m_1$

$[T_3 = 8.9 \text{ MeV}, T_4 = 129.4 \text{ MeV}]$

4.12 A particle with a non-zero rest mass m is moving in the x-direction with a speed that gives a Lorentz factor γ. The energy is then $\mathcal{E} = \gamma mc^2$. What is the momentum in the x-direction? (γmv)

4.13 The mass of some nuclear fuel is 150 g. If the mass of the nuclear fuel left after nuclear reaction is 95.0 g, how much energy is released. $c = 3.00 \times 10^8$ m/s. ($E = 4.95 \times 10^{15}$ J)

4.14 Show that when **f** is parallel to **v**,

$$\mathbf{f} = m(1 - v^2/c^2)^{-3/2} d\mathbf{v}/dt$$

where m is the mass of the object and **v** is its speed.

4.15 According to the β-decay theory, a free neutron is known to decay into a proton, an electron, and an anti-neutrino (of zero rest mass)

$$n \rightarrow p + e^- + \bar{\nu}$$

Calculate the total kinetic energy released in the decay. (0.7823 MeV)
Given:

$m_n = 939.5656$ MeV/c^2

$m_p = 938.2723$ MeV/c^2

$m_c = 0.510$ MeV/c^2

4.16 The K^{0-} meson decays into two charged pions according to the reaction

$$K^0 \rightarrow \pi^- + \pi$$

The pions have equal and opposite charges as shown and the same rest mass 140 MeV/c^2.

A K^0 at rest decays into two pions in a bubble chamber in which a magnetic field $B = 2.0$ T is applied. If the radius of curvature of the pions is 34.4 cm, determine the momenta and the speeds of pions and the rest mass of K^0 meson.

[206 MeV/c, $v = 0.827c$, and $E = 249$ MeV for each pion. $m_{k0} = 498$ MeV/c^2]

4.17 Find the momentum p of a proton $(m_0 = 938 \text{ MeV}/c^2)$ moving with kinetic energy of 1,000 MeV.
$[\text{Hint: } pc = T^2 + 2m_0T]$ (1696 MeV/c)

4.18 At what value of γ and β, the kinetic energy T of an electron is four times its rest mass energy. Rest energy of electron = 0.51 MeV. ($\gamma = 5$; $\beta = 0.9798$)

4.19 Find the threshold energy for anti-proton in the reaction

$$p + p = p + p + p + \bar{p}$$

where the struck proton inside the nucleus has kinetic energy 20 MeV. (4.6 BeV)

5

Optical Applications of Lorentz Transformation

5.1 ELEMENTS OF WAVE PROPAGATION

Let us consider a function $f_1(x, t)$ defined by the relation

$$f_1(x, t) \equiv a \sin (bx - gt) \tag{5.1}$$

where a, b, and g are all positive. The function is periodic both in x and in t and varies between $\pm a$. It can be shown to represent a wave of amplitude a, propagating in the direction of $+x$ with velocity g/b. If x is increased by δx and t by δt then

$$\begin{aligned} f_1(x + \delta x, t + \delta t) &= a \sin [b(x + \delta x) - g(t + \delta t)] \\ &= a \sin [(bx - gt) + (b\delta x - g\delta t)] \end{aligned}$$

Now,

$$f_1(x + \delta x, t + \delta t) = f_1(x, t)$$

provided

$$b\delta x - g\delta t = 0$$

or

$$\frac{\delta x}{\delta t} = \frac{g}{b} \equiv v_\omega \tag{5.2}$$

where, v_ω is called the phase velocity. The function $f_1(x, t)$ goes through one full period when x is increased by $2\pi/b$ which we call the wavelength. We denote it by λ, so that

$$\frac{2\pi}{b} \equiv \lambda \tag{5.3}$$

The function also goes through one period when keeping x constant, t is increased by $2\pi/g$. This we call the time period so that

$$\frac{2\pi}{g} \equiv \tau \tag{5.4}$$

and frequency

$$\nu = \frac{1}{\tau} = \frac{g}{2\pi} \tag{5.5}$$

Multiplying Eqs (5.3) and (5.4), we get

$$v\lambda = \frac{g}{2\pi}\frac{2\pi}{b} = \frac{g}{b} = v_\omega$$

Similar considerations hold also for the function $a \sin(bx + gt)$ with the only difference that it represents a wave travelling in the negative direction of the x-axis.

Consider a second wave with the same amplitude and slightly different wavelength and frequency

$$f_2(x, t) = a \sin[(b + \delta b)x - (g + \delta g)t]$$

and find its resultant with $f_1(x, t)$, then

$$\begin{aligned} f_1(x, t) + f_2(x, t) &= a[\sin(bx - gt) + \sin\{(b + \delta b)x - (g + \delta g)t\}] \\ &= 2a \sin[(b + \delta b/2)x - (g + \delta g/2)t] \cos\frac{1}{2}(x\delta x - t\delta g) \\ &= 2a \cos\frac{1}{2}(x\delta b - t\delta g) \sin(bx - gt) \text{ approximately} \end{aligned}$$

The resultant represents a wave motion approximately with the phase velocity $\mathbf{v}_\omega$ but with varying amplitude $2a \cos\frac{1}{2}(x\delta b - t\delta g)$. The variation of amplitude is propagated with velocity

$$v_g = \frac{dg}{db} = \frac{dv}{d\left(\frac{1}{\lambda}\right)} \quad [5.5(a)]$$

v_g is called the group velocity of the waves and gives the rate at which energy is being propagated.

$$v_g = \frac{dv}{d\left(\frac{1}{\lambda}\right)} = \frac{d\left(\frac{v_\omega}{\lambda}\right)}{d\left(\frac{1}{\lambda}\right)} = v_\omega + \frac{1}{\lambda}\frac{dv_\omega}{d\left(\frac{1}{\lambda}\right)}$$

$$= v_\omega - \lambda\frac{dv_\omega}{d\lambda} \quad [5.5(b)]$$

The group velocity differs from phase velocity only when there is wavelength dependence of velocity.

Extending the above case of one-dimensional motion to three-dimensional propagation, where the wave travels in a direction making constant angles, whose direction cosines are (l, m, n), with the coordinate axes. Then,

$$f(x, y, z, t) = a \sin 2\pi\left(vt - \frac{lx + my + nz}{\lambda}\right) \quad (5.6)$$

The wave amplitude $f(x, y, z, t)$ has the same value when the argument $\left(vt - \frac{lx + my + nz}{\lambda}\right)$ has the constant value or

$$v_\omega t - lx - my - nz = \text{constant}$$

This represents the equation of a plane perpendicular to (l, m, n) and moving with velocity v_ω, so we infer that Eq. (5.6) is a plane wave solution of the wave equation

$$\left(\nabla^2 - \frac{1}{v_\omega^2}\frac{\partial^2}{\partial t^2}\right) f(t, x, y, z) = 0$$

The electromagnetic equations expressed in terms of vector potential **A** and scalar potential ϕ have the form

$$\Box^2 \mathbf{A} = -\frac{4\pi}{\mathbf{c}}\rho \mathbf{u}$$

$$\Box^2 \phi = -4\pi\rho$$

When there are no charges in the electromagnetic field, $\rho = 0$ and the above equations reduce to

$$\left(\nabla^2 - \frac{1}{v_\omega^2}\frac{\partial^2}{\partial t^2}\right) \mathbf{A}(t, x, y, z) = 0 \tag{5.7}$$

$$\left(\nabla^2 - \frac{1}{v_\omega^2}\frac{\partial^2}{\partial t^2}\right) \phi(t, x, y, z) = 0 \tag{5.8}$$

Obviously **A** and ϕ satisfy the same equation.
Rewriting the equations for **B** and **E**

$$\mathbf{B} = \nabla \times \mathbf{A}$$

$$\mathbf{E} = -\frac{1}{c}\frac{\partial \mathbf{A}}{\partial t} - \nabla\phi$$

Making use of the above equations for **B** and **E** and Eqs (5.7) and (5.8), we get

$$\left(\nabla^2 - \frac{1}{c^2}\frac{\partial^2}{\partial t^2}\right) \mathbf{B} = 0 \tag{5.9}$$

$$\left(\nabla^2 - \frac{1}{c^2}\frac{\partial^2}{\partial t^2}\right) \mathbf{E} = 0 \tag{5.10}$$

Thus we infer that for a plane wave solution, the electromagnetic field strengths are propagated with velocity c. Since c is independent of wavelength, the group and phase velocities are similar for electromagnetic waves.

It is more customary to represent a wave motion by the real or imaginary part of the expression

$$a\ exp\ 2\pi i\left(vt - \frac{lx + my + nz}{\lambda}\right) \tag{5.11}$$

5.2 FREQUENCY AND WAVE-VECTOR TRANSFORMATIONS

We here bring out the transformation properties of frequency and wave vector components of an electromagnetic wave. The wave vector potential **A** of a monochromatic plane wave is given by

$$\mathbf{A} = \mathrm{Re}\left[\mathrm{A}_0 e^{i(\omega t - \mathbf{k.r})}\right]$$

The phase of a wave is an invariant quantity. Since the phase characterizes some definite event such as the fading of the electric and magnetic fields at a certain instant of time and at a certain point in space, when this event is measured in another coordinate system, the coordinates and time corresponding to this event will have transformed values, but the event itself must be the same. Were it not the case, we will be in a position to find a coordinate system which will be unique and distinguishable from the others. But this will be against the principle of relativity.

The phase of the wave denoted by ψ is given by

$$\psi = \omega t - xk_x - yk_y - zk_z$$

The invariance of phase can be expressed as

$$\omega t - xk_x - yk_y - zk_z = \omega' t' - x'k'_x - y'k'_y - z'k'_z$$

$$= \omega' \frac{t - vx/c^2}{\sqrt{1 - v^2/c^2}} - \frac{x - vt}{\sqrt{1 - v^2/c^2}} k'_x - y'k'_y - z'k'_z$$

Comparing the coefficients of t, x, y and z, we obtain the following transformation formulae for frequency and the wave vector components

$$\omega = \frac{\omega' + k'_x v}{\sqrt{1 - v^2/c^2}} \tag{5.12}$$

$$k_x = \frac{k'_x + \omega' v}{c^2/\sqrt{1 - v^2/c^2}} \tag{5.13}$$

$$k_y = k'_y \tag{5.14}$$

$$k_z = k'_z \tag{5.15}$$

We infer that the set of four quantities $(\omega/c, \mathbf{k}) = k_\mu$ is a wave four-vector.

5.3 THE RELATIVISTIC DOPPLER EFFECT. ABERRATION

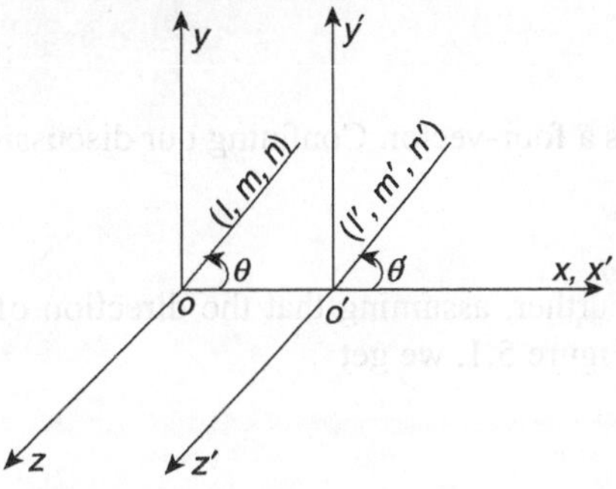

Figure 5.1 *A plane wave in two Lorentz frames.*

Suppose that a plane wave given by

$$a\ exp\ 2\pi i\left(vt - \frac{lx + my + nz}{\lambda}\right) \tag{5.16}$$

is sent out by the observer B from the primed axes and observed by the observer A in the unprimed axes, Figure 5.1. The plane wave with reference to the primed axes is represented by

$$a\ exp\ 2\pi i\left(v't' - \frac{l'x' + m'y' + n'z'}{\lambda'}\right) \tag{5.17}$$

The argument of the exponential functions must be same, since these represent the same wave. Moreover, these cannot differ by a multiple of 2π, since these must be the same for identical transformations

$$x' = x = 0$$

$$y' = y = 0$$

$$z' = z = 0$$

$$t' = t = 0$$

Expressing x', y', z' and t' in terms of x, y, z, and t, we get

$$a\ exp\ 2\pi i\left(v'\gamma\left(t - \frac{vx}{c^2}\right) - \frac{l'\gamma(x - vt) + m'y + n'z}{\lambda'}\right) \tag{5.17}$$

and comparing it with Eq. (5.16), for the coefficients of x, y, z, and t, we obtain

$$v = \gamma\left(v' + \frac{vl'}{\lambda'}\right) \tag{5.18(a)}$$

$$\frac{l}{\lambda} = \gamma\left(\frac{l'}{\lambda'} + \frac{vv'}{c^2}\right) \tag{5.18(b)}$$

$$\frac{m}{\lambda} = \frac{m'}{\lambda'} \tag{5.18(c)}$$

$$\frac{n}{\lambda} = \frac{n'}{\lambda'} \tag{5.18(d)}$$

We infer from Eqs (5.18) that

$$\frac{v}{c}, \frac{l}{\lambda}, \frac{m}{\lambda}, \frac{n}{\lambda}$$

is a four-vector. Confining our discussion to the electromagnetic waves, we have

$$v\lambda = v'\lambda' = c$$

Further, assuming that the direction of propagation makes angles θ and θ' with the x- and x'-axes, Figure 5.1, we get

$$l = \cos\theta;\ l' = \cos\theta'$$

and from Eqs (5.18)

$$\frac{l}{\lambda} = \gamma\left(\frac{l'}{\lambda'} + \frac{vv'}{c^2}\right)$$

or

$$\frac{lv}{c} = \gamma\left(\frac{l'v'}{c} + \frac{v}{c\lambda'}\right)$$

But

$$\frac{lv}{c} = \gamma\left(v' + \frac{vl'}{\lambda'}\right)$$

so

$$\frac{l}{c}\gamma\left(v' + \frac{vl'}{\lambda'}\right) = \gamma\left(\frac{l'v'}{c} + \frac{v}{c\lambda'}\right)$$

which leads to

$$\frac{l}{c} = \frac{l' + v/c}{v'\lambda' + vl'} = \frac{l' + v/c}{c + vl'}$$

Expressing it in terms of angles

$$\cos\theta = \frac{\cos\theta' + v/c}{1 + v\cos\theta'/c} \tag{5.19}$$

Rewriting the following equation from Eqs (5.18)

$$v = \gamma\left(v' + \frac{vl'}{\lambda'}\right)$$

or

$$\frac{v}{v'} = \frac{\lambda'}{\lambda} = \gamma\left(1 + \frac{v}{c}\cos\theta'\right) \tag{5.20}$$

Equation (5.19) gives aberration of light, and Eq. (5.20), the relativistic Doppler effect of light, i.e., the apparent frequency of a light source which is moving uniformly relative to the observer A. Two special cases arise.

5.3.1 The Longitudinal Doppler Effect

When source B is receding directly from observer A along the x-axis, we have $\theta = \theta' = 0$ and hence

$$\nu = \nu' \sqrt{\frac{1 + v/c}{1 - v/c}} \tag{5.21}$$

On the contrary, if the source moves along the line of sight, then $\theta = \pi$ and hence

$$\nu = \frac{\nu'(1 - v/c)}{\sqrt{1 - v^2/c^2}} = \nu' \sqrt{\frac{1 - v/c}{\sqrt{1 + v/c}}}$$

or

$$\frac{\nu}{\nu'} = \frac{\lambda'}{\lambda} = \sqrt{\frac{1 - v/c}{\sqrt{1 + v/c}}} \tag{5.22}$$

Equation (5.22) is employed by astronomers to interpret *the red shift of spectral lines* of light received from distant nebulae. If λ' is the wavelength as emitted by the source and λ that as detected on earth, the red shift implies that, $\lambda > \lambda'$, which will be true only when v is positive. Thus the nebulae and galaxies appear to be receding from us. This is taken as evidence in favour of the expanding universe and the Big Bang theory.

5.3.2 The Transverse Doppler Effect

When the velocity of source is perpendicular to the direction of propagation, then $\theta' = \pi/2$ and Eq. (5.20) becomes

$$\nu = \frac{\nu'}{\sqrt{1 - v^2/c^2}} \tag{5.23}$$

A change of frequency occurs even when the observer is looking at right angles to the direction of motion of the source. No such effect is predicted classically. The transverse Doppler effect was verified experimentally by Ives and Stilwell[1] in the case of radiation from moving ions in canal rays by noting the shift in frequency spectroscopically.

Example 5.1 Show that Eq. (5.19) for the aberration of light can be written as

$$\sin\theta = \frac{\sqrt{1 - v^2/c^2}\,\sin\theta'}{1 + v\cos\theta'/c}$$

Solution. From Eq. (5.19), we have

$$(1 + \cos\theta) = 1 + \frac{\cos\theta' + v/c}{1 + v\cos\theta'/c}$$

$$= \frac{(1 + \cos\theta')\,(1 + v/c)}{1 + v\cos\theta'/c}$$

$$(1 - \cos\theta) = \frac{(1 - \cos\theta')\,(1 - v/c)}{1 + v\cos\theta'/c}$$

Multiplying the above two equations

$$(1-\cos^2\theta) = \frac{(1-\cos^2\theta')(1-v^2/c^2)}{(1+v\cos\theta'/c)^2}$$

or

$$\sin\theta = \frac{\sqrt{1-v^2/c^2}\,\sin\theta'}{1+v\cos\theta'/c} \tag{5.24}$$

Example 5.2 Derive the transformation formula for a solid angle element in spherical coordinates.

Solution. Orient the polar axis in the direction of relative motion of both the frames (the x-, x'-axes). In the primed frame the solid angle element $d\Omega'$ is

$$d\Omega' = \sin\theta'\,d\theta'\,d\varphi' = -d(\cos\theta')\,d\varphi'$$

Since the y and z coordinates do not change, the φ coordinate, that is the projection on a plane perpendicular to the motion direction, does not change either: $\varphi = \varphi'$ and $d\varphi = d\varphi'$. From the aberration formula

$$\cos\theta' = \frac{\cos\theta-\beta}{1-\beta\cos\theta}$$

we get

$$d(\cos\theta') = -\frac{1-\beta^2}{(1-\beta\cos\theta)^2}\sin\theta d\theta$$

Therefore,

$$d\Omega' = \frac{d\Omega}{\gamma^2(1-\beta\cos\theta)^2}\sin\theta d\theta\,d\phi = \frac{d\Omega}{\gamma^2(1-\beta\cos\theta)^2}$$

since in the unprimed frame, the solid angle element $d\Omega$ is equal to

$$d\Omega = \sin\theta d\theta\,d\varphi$$

Example 5.3 Light emitted from a source in S' is limited by stops, which obstruct all the light except that within a small cone of solid angle $d\Omega'$ (measured in S'). Show that the solid angle measured by an observer in a system S, which is approaching the source with velocity v, is given by

$$\frac{d\Omega}{d\Omega'} = \frac{1-v/c}{1+v/c}$$

Solution. Rewriting Eq. (5.24)

$$\sin\theta = \frac{\sqrt{1-v^2/c^2}\,\sin\theta'}{1+v\cos\theta'/c}$$

Differentiating it, we get

$$\cos\theta d\theta = \left[\frac{\sqrt{1-v^2/c^2}\,\cos\theta'}{1+v\cos\theta'/c} + v/c\,\frac{\sqrt{1-v^2/c^2}\,\sin^2\theta'}{(1+v\cos\theta'/c)^2}\right]d\theta'$$

When $\theta = \theta' = 0$, we have

$$d\theta = \frac{\sqrt{1 - v^2/c^2}}{1 + v/c}\, d\theta'$$

or

$$\frac{d\theta}{d\theta'} = \frac{\sqrt{1 - v^2/c^2}}{1 + v/c}$$

But

$$\frac{d\Omega}{d\Omega'} = \frac{\sin\theta d\theta}{\sin\theta' d\theta'} = \frac{1 - v^2/c^2}{(1 + v/c)(1 + v/c)} = \frac{1 - v/c}{1 + v/c}$$

This gives the change in the solid angle of a pencil of radiation due to motion of the observer in the line of sight, the source approaching the observer with velocity v. However, if the source is receding with respect to the observer with velocity v, then

$$\frac{d\Omega}{d\Omega'} = \frac{1 + v/c}{1 - v/c} \tag{5.25}$$

Example 5.4 *Intensity of a moving source:* There is a point source of light P, which is moving directly away from an observer A, along ox, with velocity v. Estimate the intensity that A will observe.

Solution. As compared with a fixed source, the intensity estimated by A is diminished by the following facts:

1. The energy of each light quantum $h\nu$ observed is reduced by the Doppler effect in the ratio given by Eq. (5.22)

$$\frac{\nu}{\nu'} = \frac{\lambda'}{\lambda} = \sqrt{\frac{1 - v/c}{1 + v/c}}$$

2. The actual number of quanta falling per unit area per second at A is reduced in the same ratio of

$$\sqrt{\frac{1 - v/c}{1 + v/c}}$$

and solid angle in which these quanta are emitted appears to A as

$$\frac{d\Omega}{d\Omega'} = \frac{1 + v/c}{1 - v/c}$$

Now the intensity is defined as the energy flux falling per unit time on unit area normal to the direction of motion at A. This is proportional to

$$\frac{\text{Energy per quantum} \times \text{number of quanta arriving per second}}{\text{solid angle subtended by unit area at the source}}$$

Thus we get the expression for intensity as

$$\frac{\text{constant}}{d^2}\left(\frac{1 - v/c}{1 + v/c}\right)^2 \tag{5.26}$$

where d is the distance of P from A at the instant of emission of light. The constant depends only upon the source and not on its motion, and is the energy per unit area at unit distance measured by an observer moving with the source P.

5.4 REFLECTION OF LIGHT BY A MOVING MIRROR

A mirror is moving in the direction of its own normal with a velocity **v** relative to an observer and the source, and we intend to find the law of reflection of light at it. Let S be the system in which mirror is stationary and S' the system of the observer and χ be the angle of incidence measured in S. Then the angle of reflection measured in S will be $-\chi$, since mirror is stationary in S and the ordinary law of reflection holds in it. Let χ'_i and χ'_r be the angles of incidence and reflection measured in S', Figure 5.2.

We get the relation between χ and χ'_i making use of Eq. (5.24), thus

$$\sin\chi = \frac{\sqrt{1-v^2/c^2}\,\sin\chi'_i}{1+v\cos\chi'_i/c} \tag{5.27}$$

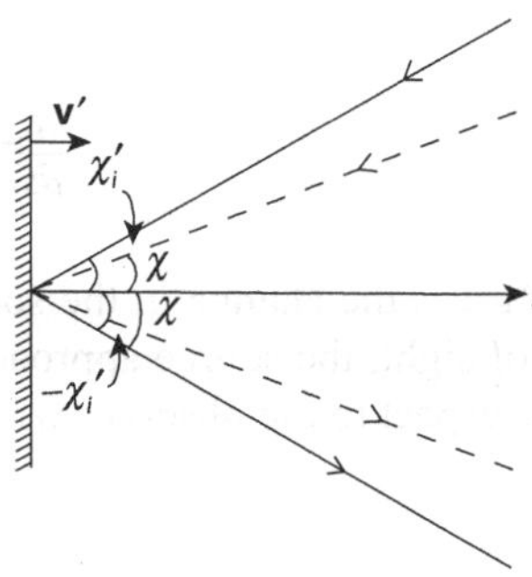

Figure 5.2 *Reflection at a moving mirror.*

Further, employing Eq. (5.27) again, we seek the relation between χ and χ'_i. Changing the sign of v, because the incident beam travels towards the mirror and the reflected beam away from it, we have

$$\sin\chi = \frac{\sqrt{1-v^2/c^2}\,\sin\chi'_r}{1-v\cos\chi'_r/c}$$

or

$$\sin(-\chi) = -\frac{\sqrt{1-v^2/c^2}\,\sin\chi'_r}{1-v\cos\chi'_r/c} \tag{5.28}$$

From Eqs (5.27) and (5.28), we get the law of reflection as

$$\frac{\sin\chi'_i}{1+v\cos\chi'_i/c} = -\frac{\sin\chi'_r}{1-v\cos\chi'_r/c} \tag{5.29}$$

5.5 PROPAGATION OF LIGHT IN A MOVING MEDIUM: FRESNEL'S DRAGGING COEFFICIENT[2]

Consider a transparent medium, such as glass or water, at rest in a frame S' but moving with velocity v relative to another frame S. If u' is the velocity of light in medium relative to S' and $n(\lambda')$ is the refractive index for vacuum wavelength λ', then

$$u' = c/n(\lambda') \tag{5.30}$$

Our problem is to find the velocity u of light in this moving medium as measured by observer in S. The wavelength in vacuum as measured by S is λ. Now from the law of composition of velocities [Eq. (2.72)] we have

$$u = \frac{u' + v}{1 + \dfrac{u'v}{c^2}} \tag{5.31}$$

Substituting Eq. (5.30) in the above equation, we obtain

$$u = \frac{\dfrac{c}{n(\lambda')} + v}{1 + \dfrac{v}{cn(\lambda')}} \tag{5.32}$$

where $n(\lambda')$ is the refractive index of the medium for the vacuum wavelength λ'. But we need a formula involving λ instead of λ'. From Eq. (5.21)

$$\nu = \nu' \sqrt{\frac{1 + v/c}{1 - v/c}} \tag{5.33}$$

or since $c = \nu\lambda = \nu'\lambda'$, we get

$$\frac{\lambda'}{\lambda} = \frac{\nu}{\nu'} = \sqrt{\frac{1 + v/c}{1 - v/c}} \tag{5.34}$$

Thus the exact expression for the velocity u of light of (vacuum) wavelength λ in a medium of refractive index $n(\lambda)$, moving with velocity v is given by, from Eq. (5.32),

$$\frac{u}{c} = \frac{\dfrac{1}{n(\lambda\xi)} + \dfrac{v}{c}}{1 + \dfrac{v}{cn(\lambda\xi)}}; \quad \xi = \sqrt{\frac{1 + v/c}{1 - v/c}} \tag{5.35}$$

When the ratio v/c is small, we may use an approximation and so

$$\lambda\xi = \lambda + \lambda\frac{v}{c} \tag{5.36}$$

and hence

$$\frac{1}{n(\lambda\xi)} = \frac{1}{n(\lambda)} + \frac{v}{c}\frac{d}{d\lambda}\frac{1}{n(\lambda)} + \cdots \tag{5.37}$$

Substituting Eq. (5.37) into Eq. (5.35) and neglecting v^2/c^2, we get

$$\frac{u}{c} = \frac{\dfrac{1}{n(\lambda)} + \lambda\dfrac{v}{c}\dfrac{d}{d\lambda}\dfrac{1}{n(\lambda)} + \dfrac{v}{c}}{1 + \dfrac{v}{c}\dfrac{1}{n(\lambda)}}$$

$$= \frac{1}{n(\lambda)} + \frac{v}{c}\left[1 - \frac{1}{[n(\lambda)]^2} + \lambda \frac{d}{d\lambda}\frac{1}{n(\lambda)}\right] \tag{5.38}$$

or

$$u = \frac{c}{n(\lambda)} + Kv$$

where

$$K = 1 - \frac{1}{[n(\lambda)]^2} - \frac{1}{[n(\lambda)]^2}\frac{d}{d\lambda}n(\lambda) \tag{5.39}$$

The quantity K is called Fresnel's drag coefficient.
If the medium is non-dispersive, that is, $(d/d\lambda)\, n(\lambda) = 0$, then

$$K = 1 - \frac{1}{[n(\lambda)]^2}$$

and

$$u = \frac{c}{n} + Kv \tag{5.40}$$

Equation (5.40) was tested and confirmed by Zeeman's repetition of Fizeau's experiment, which compares the velocity of light in opposite directions in a moving stream of water. Water is not highly dispersive, since the refractive index n of water at 20°C varies only from 1.3308 to 1.3428 as the wavelength changes from 6708 Å to 4047 Å for range of visible spectrum and hence the formula is expected to agree with experiment, and so it does. This is a verification of the relativity law of composition of velocities. No assumption of any dragging is involved in our argument and hence no postulation of an ether is made.

References

1. H. E. Ives and G. R. Stilwell "An Experimental Study of the Rate of a Moving Clock", *J. Opt. Soc. Am.* (1938) *28*, 215; (1941) *31*, 369.
2. L. Blitzer "On the Meaning of the Fresnel Coefficient of Ether Drag in Relativity", *Am. J. Phys.* (1947) *15*, 446.

Problems

5.1 Show that the function $f_1(x, t) = a \sin(bx - gt)$ representing a wave, satisfies the wave equation

$$\frac{\partial^2 f_2(x,t)}{\partial x^2} - \frac{1}{v_\omega^2}\frac{\partial^2 f_1(x,t)}{\partial t^2}\, 0$$

where v_ω is the phase velocity of the wave.

5.2 Prove that the absolute value of the field of a monochromatic wave transforms under Lorentz transformation like the first power of frequency.

5.3 If θ is the angle between the direction of the light ray and the direction of relative motion of S and S', show the dependence of the frequency of light ray ν' on θ. Derive both the classical and the relativistic equations showing how ν' depends on ν and the angle θ.

5.4 A certain source emits light of wavelength 6,000 Å, measured by an observer for whom the source is stationary. Find the displacements (a) for an observer who observes a transverse motion of the source, and (b) for an observer who observes a recession at 89° 55′ to the line of sight. In each case the velocity is 0.0031 c.

5.5 Show that if ω'_i and ω'_r are the circular frequencies of the incident and reflected radiation, respectively, then

$$\omega'_i\left(1+\tfrac{v}{c}\cos\chi'_i\right)=\omega'_r\left(1-\tfrac{v}{c}\cos\chi'_r\right)$$

Hint: If ν is the frequency of the incident (and also of the reflected) radiation in **S**, then the relation between ν and ν' is given by

$$\frac{\nu}{\nu'}=\gamma\left(1+\tfrac{v}{c}\cos\theta'\right) \tag{5.20}$$

Making use of Eq. (5.20), get the relations between ω and ω'_i and between ω and ω'_r.

5.6 A point source of light is placed at the focus of a lens whose focal length is f and a plane mirror is set normal to the axis of the lens so as to give an image which coincides with the object when the mirror is stationary. Show that when the mirror moves with speed v there is a change of wavelength proportional to v^2/c^2.

Hint: Make use of Eq. (5.20).

5.7 The refractive index of carbon disulphide for D lines of sodium is $n_D = 1.6295$ and $dn/d\lambda = -1{,}820$ per cm at this wavelength. Find (a) the ratio of the velocity of light in vacuum to the group velocity in carbon disulphide and (b) the value of Fresnel's dragging coefficient.

5.8 Calculate the wavelength shift in the relativistic Doppler effect for the $\mathcal{H}\alpha$ [6,563 Å] line emitted by star receding from earth with a relative velocity 0.1 c. (692.34 Å)

5.9 The wavelength of spectral line is $\lambda_0 = 4{,}861$ Å. Estimate longitudinal and transverse red shifts when the atom is moving with energy 6, 780 e V. (18.2 Å; 0.035 Å)

5.10 A pencil of light rays is emitted into a solid angle $d\Omega'$ inclined at an angle θ' to the x' axis in S'. Show that in S which moves in the negative x'-direction with uniform velocity v relative to S', the pencil of light rays is emitted into a solid angle $d\Omega'$ given by

$$d\Omega=\frac{(1-\beta^2)}{(1+\beta\cos\theta')^2}\,d\Omega' \quad \text{where } \beta=\tfrac{v}{c}.$$

Show that the solid angle $d\Omega$ is inclined at an angle θ to the x-axis in S, where

$$\cos\theta=\frac{\cos\theta'+\beta}{1+\beta\cos\theta'}$$

5.11 For wavelength Å, find longitudinal and transverse red shifts when the source is moving with $\beta = 0.5$. (4392 Å, 930 Å)

5.12 A distant star in the direction perpendicular to the plane of earth's orbit describes circular annual orbit in angular diameter. Find the speed v of the earth from the stellar aberration data.

Hint: $\sin \phi = v/c$. (30 km/s)

5.13 (a) Find the speed and the direction in which galaxy A must be moving if an absorption line found at wavelength 550 nm (green) for a stationary galaxy is shifted to 450 nm (blue) (a blue-shift) for galaxy A. (0.198c)

(b) At what speed and in what direction a galaxy B must be moving if it shows the same line shifted to 700 nm(red) (a red-shift). (0.237c)

6

Covariant Electrodynamics

6.1 INVARIANCE OF MAXWELL'S EQUATIONS UNDER LORENTZ TRANSFORMATION

It is shown in Section 1.4 that the wave equations

$$\nabla^2 \mathbf{B} - \frac{1}{c^2}\frac{\partial^2 \mathbf{B}}{\partial t^2} = 0 \tag{6.1}$$

$$\nabla^2 \mathbf{E} - \frac{1}{c^2}\frac{\partial^2 \mathbf{E}}{\partial t^2} = 0 \tag{6.2}$$

are not invariant under Galilian transformation. Our immediate objective is to prove that Lorentz transformations also leave Maxwell–Lorentz equations invariant.

Maxwell–Lorentz equations governing the field created by a given distribution of charge and current are (in Gaussian units in which the field vectors have the same dimensions)

$$\frac{1}{c}\frac{\partial \mathbf{E}}{\partial t} + \frac{4\pi}{c}\rho \mathbf{u} = \nabla \times \mathbf{B} \tag{6.3}$$

$$\nabla \times \mathbf{E} + \frac{1}{c}\frac{\partial \mathbf{B}}{\partial t} = 0 \tag{6.4}$$

$$\nabla \cdot \mathbf{E} = 4\pi\rho \tag{6.5}$$

$$\nabla \cdot \mathbf{B} = 0 \tag{6.6}$$

where **E** and **B** are respectively the electric and magnetic intensities, c is the ratio of electromagnetic and electrostatic units of charge, ρ is the density of electrical charge, **u** is its velocity, and $\rho\mathbf{u}$ is the current density.

Let us now introduce the scalar potential ϕ and vector potential **A**. From Eq. (6.6), we deduce that **B** can always be represented as the curl of another vector **A**, which we call the vector potential.

$$\mathbf{B} = \nabla \times \mathbf{A} \tag{6.7}$$

Thus, Eq. (6.4) becomes

$$\nabla \times \mathbf{E} = -\frac{1}{c}\frac{\partial \mathbf{B}}{\partial t} = -\frac{1}{c}\nabla \times \frac{\partial \mathbf{A}}{\partial t}$$

Therefore,

$$\mathbf{E} = -\frac{1}{c}\frac{\partial \mathbf{A}}{\partial t} - \nabla\phi \tag{6.8}$$

since the curl of a gradient vanishes.

Rewriting Eq. (6.5), we have

$$4\pi\rho = \nabla\cdot\mathbf{E}$$

$$= \nabla\cdot\left[-\frac{1}{c}\frac{\partial\mathbf{A}}{\partial t} - \nabla\phi\right]$$

$$= -\frac{1}{c}\frac{\partial}{\partial t}(\nabla\cdot\mathbf{A}) - \nabla^2\phi$$

or

$$\Box^2\phi + \frac{1}{c}\frac{\partial}{\partial t}\left(\nabla\cdot\mathbf{A} + \frac{1}{c}\frac{\partial\phi}{\partial t}\right) = -4\pi\rho \tag{6.9}$$

where

$$\Box^2 \equiv \frac{\partial^2}{\partial x^2} + \frac{\partial^2}{\partial y^2} + \frac{\partial^2}{\partial z^2} - \frac{1}{c^2}\frac{\partial^2}{\partial t^2} = \nabla^2 - \frac{1}{c^2}\frac{\partial^2}{\partial t^2} \tag{6.10}$$

is d′ Alembertian operator. Rewriting Eq. (6.3), we get

$$\frac{1}{c}\frac{\partial\mathbf{E}}{\partial t} + \frac{4\pi}{c}\rho\mathbf{u} = \nabla\times\mathbf{B}$$

$$\frac{1}{c}\frac{\partial}{\partial t}\left(-\frac{1}{c}\frac{\partial\mathbf{A}}{\partial t} - \nabla\phi\right) + \frac{4\pi}{c}\rho\mathbf{u} = \nabla\times\nabla\times\mathbf{A}$$

$$\frac{4\pi}{c}\rho\mathbf{u} = \nabla\times\nabla\times\mathbf{A} - \frac{1}{c}\frac{\partial}{\partial t}\left[-\frac{1}{c}\frac{\partial\mathbf{A}}{\partial t} - \nabla\phi\right]$$

$$= -\nabla^2\mathbf{A} + \nabla(\nabla\cdot\mathbf{A}) + \frac{1}{c^2}\frac{\partial^2\mathbf{A}}{\partial t^2} + \frac{1}{c}\nabla\frac{\partial\phi}{\partial t}$$

that is,

$$\Box^2\mathbf{A} - \nabla\left(\nabla\cdot\mathbf{A} + \frac{1}{c}\frac{\partial\phi}{\partial t}\right) = -\frac{4\pi}{c}\rho\mathbf{u} \tag{6.11}$$

The field equations have been reduced to a simpler form of equations between a scalar and a vector, instead of two vectors.

The vector **A** is not determined completely by the magnetic field **B**. Since for any scalar function χ curl grad $\chi = 0$, we can add to **A** the gradient of an arbitrary scalar function χ. According to the equation $\mathcal{E} = -\frac{1}{c}\frac{\partial\mathbf{A}}{\partial t} - \nabla\phi$, we have to replace ϕ by $\phi + \frac{1}{c}\frac{\partial\chi}{\partial t}$ if we replace **A** by $\mathbf{A} - \nabla\chi$ so as to preserve $\mathcal{E}$. This freedom in the choice of potentials, called gauge transformation, can be used to simplify the field equations. Calling A_0, ϕ_0 one set of values of **A**, ϕ, we determine χ from the equation

$$\nabla\cdot\mathbf{A} + \frac{1}{c}\frac{\partial\phi}{\partial t} = \nabla\cdot(\mathbf{A}_0 - \nabla\chi) + \frac{1}{c}\frac{\partial}{\partial t}\left(\phi_0 + \frac{1}{c}\frac{\partial\chi}{\partial t}\right)$$

$$= \nabla\cdot\mathbf{A}_0 + \frac{1}{c}\frac{\partial\phi_0}{\partial t} - \left(\nabla^2\chi - \frac{1}{c^2}\frac{\partial^2\chi}{\partial t^2}\right) \tag{6.12}$$

Choosing χ to satisfy the equation

$$\Box^2\chi = \nabla\cdot\mathbf{A}_0 + \frac{1}{c}\frac{\partial\phi_0}{\partial t} \tag{6.13}$$

We get

$$\Box^2\mathbf{A} = -\frac{4\pi}{c}\rho\mathbf{u} \tag{6.14}$$

$$\Box^2\phi = -4\pi\rho \tag{6.15}$$

$$\nabla\cdot\mathbf{A} + \frac{1}{c}\frac{\partial\phi}{\partial t} = 0 \tag{6.16}$$

Equation (6.16) represents a relation between the potentials and is called the *Lorentz relation*. The field equations reduce to the above inhomogeneous equations, satisfied by **A** and ϕ respectively.

The invariance of the field equations can be established if Eqs [(6.14)–(6.16)] preserve their form under the transformation of the four-potential and four-current. Both the equations have an operator $\nabla^2 - \frac{1}{c^2}\frac{\partial}{\partial t^2}$ on the left-hand side, which is invariant under Lorentz transformation.

The transformations of the different components of four-potential $A^\mu(\phi, \mathbf{A})$ and four-current $j^\mu(c\rho, \mathbf{J})$ are as follows:

$$\phi' = \gamma\left(\phi - \frac{v}{c}A_x\right) \quad [6.17(a)]$$

$$A'_x = \gamma\left(A_x - \frac{v}{c}\phi\right) \quad [6.17(b)]$$

$$A'_y = A_y \quad [6.17(c)]$$

$$A'_z = A_z \quad [6.17(d)]$$

and

$$\rho' = \gamma\left(\rho - \frac{v}{c^2}J_x\right) \quad [6.18(a)]$$

$$J'_x = \gamma(J_x - v\rho) \quad [6.18(b)]$$

$$J'_y = J_y \quad [6.18(c)]$$

$$J'_z = J_z \quad [6.18(d)]$$

$$\Box^2 A'_x = -\frac{4\pi}{c}\gamma(\rho u_x - v\rho) = -\frac{4\pi}{c}(\rho u_x)' \text{ etc.} \tag{6.19}$$

$$\Box^2\phi' = -4\pi\gamma\left(\rho - \frac{v}{c^2}\rho u_x\right) = -4\pi\rho' \tag{6.20}$$

and

$$0 = \nabla'\cdot\mathbf{A}' + \frac{1}{c}\frac{\partial\phi'}{\partial t'}$$

$$= \tfrac{1}{c}\gamma^2\left(v\frac{\partial}{\partial x} + \frac{\partial}{\partial t}\right)\left(\phi - \tfrac{v}{c}A_x\right) + \gamma^2\left(\frac{\partial}{\partial x} + \frac{v}{c^2}\frac{\partial}{\partial t}\right)\left(A_x - \tfrac{v}{c}\phi\right) + \frac{\partial A_y}{\partial y} + \frac{\partial A_z}{\partial z}$$

$$= \frac{1}{c}\frac{\partial \phi}{\partial t} + \frac{\partial A_x}{\partial x} + \frac{\partial A_y}{\partial y} + \frac{\partial A_z}{\partial z}$$

$$= \frac{1}{c}\frac{\partial \phi}{\partial t} + \nabla \cdot \mathbf{A} \tag{6.21}$$

Thus, Maxwell's equations are invariant under Lorentz transformations.

6.2 TRANSFORMATION OF ELECTROMAGNETIC FIELD INTENSITIES

The electric and magnetic field intensities are given by

$$\mathbf{E} = -\frac{1}{c}\frac{\partial \mathbf{A}}{\partial t} - \nabla\phi \tag{6.22}$$

$$\mathbf{B} = \nabla \times \mathbf{A} \tag{6.23}$$

The components of $\mathbf{E}'$ and $\mathbf{B}'$, the transformed field intensities, are given by the relations

$$E'_x = -\frac{1}{c}\frac{\partial A'_x}{\partial t'} - \frac{\partial \phi'}{\partial x'} \text{ etc.} \tag{6.24}$$

$$B'_x = \frac{\partial A'_z}{\partial y'} - \frac{\partial A'_y}{\partial z'} \text{ etc.} \tag{6.25}$$

Making use of Lorentz transformations, we have

$$\frac{\partial}{\partial t'} = \gamma\left(v\frac{\partial}{\partial x} + \frac{\partial}{\partial t}\right) \tag{6.26(a)}$$

$$\frac{\partial}{\partial x'} = \gamma\left(\frac{\partial}{\partial x} + \frac{v}{c^2}\frac{\partial}{\partial t}\right) \tag{6.26(b)}$$

$$\frac{\partial}{\partial y'} = \frac{\partial}{\partial y} \tag{6.26(c)}$$

$$\frac{\partial}{\partial z'} = \frac{\partial}{\partial z} \tag{6.26(d)}$$

The four-potential A'^{μ} has the following transformation properties:

$$\phi' = \gamma\left(\phi - \tfrac{v}{c}A_x\right) \tag{6.27(a)}$$

$$A'_x = \gamma\left(A_x - \tfrac{v}{c}\phi\right) \tag{6.27(b)}$$

$$A'_y = A_y \tag{6.27(c)}$$

$$A'_z = A_z \tag{6.27(d)}$$

Therefore,
$$E'_x = -\tfrac{1}{c}\frac{\partial A'_x}{\partial t'} - \frac{\partial \phi'}{\partial x'}$$
$$= -\tfrac{1}{c}\gamma^2\left(v\frac{\partial}{\partial x}+\frac{\partial}{\partial t}\right)\left(A_x - \tfrac{v}{c}\phi\right) - \gamma^2\left(\frac{\partial}{\partial x}+\frac{v}{c^2}\frac{\partial}{\partial t}\right)\left(\phi - \tfrac{v}{c}A_x\right)$$
$$= -\tfrac{1}{c}\frac{\partial A_x}{\partial t} - \frac{\partial \phi}{\partial x} = E_x \qquad [6.28(a)]$$
$$E'_y = -\tfrac{1}{c}\frac{\partial A'_y}{\partial t'} - \frac{\partial \phi'}{\partial y'}$$
$$= -\tfrac{1}{c}\gamma\left(v\frac{\partial}{\partial x}+\frac{\partial}{\partial t}\right)A_y - \gamma\frac{\partial}{\partial y}\left(\phi - \tfrac{v}{c}A_x\right)$$
$$= \gamma\left(E_y - \tfrac{v}{c}B_z\right) \qquad [6.28(b)]$$
$$E'_z = -\tfrac{1}{c}\frac{\partial A'_z}{\partial t'} - \frac{\partial \phi'}{\partial z'}$$
$$= -\tfrac{1}{c}\gamma\left(v\frac{\partial}{\partial x}+\frac{\partial}{\partial t}\right)A_z - \gamma\frac{\partial}{\partial z}\left(\phi - \tfrac{v}{c}A_x\right)$$
$$= \gamma\left(E_z + \tfrac{v}{c}B_y\right) \qquad [6.28(c)]$$

In contradistinction to the transformation of coordinates, it is not the longitudinal but the transverse components that are transformed in the field.
Analogously,
$$B'_x = \frac{\partial A'_z}{\partial y'} - \frac{\partial A'_y}{\partial z'}$$
$$= \frac{\partial A_z}{\partial_y} - \frac{\partial A_y}{\partial_z}$$
$$= B_x \qquad [6.29(a)]$$
Furthermore, likewise we can show that
$$B'_y = \gamma\left(B_y + \tfrac{v}{c}E_z\right) \qquad [6.29(b)]$$
$$B'_z = \gamma\left(B_z - \tfrac{v}{c}E_y\right) \qquad [6.29(c)]$$

Example 6.1 A charged particle carrying charge e is moving with velocity **u** under the influence of electromagnetic field of intensities **E** and **B**. Show that the force acting on it is given by the Lorentz force equation,
$$\mathbf{F} = e\left[\mathbf{E} + \tfrac{1}{c}(\mathbf{u}\times\mathbf{B})\right]$$

Solution. If at an instant the particle has velocity u, choose the primed axes S' in such a way that its origin O′ lies at the particle and is moving with it, the x'-axis being pointed in the direction of motion of the particle. The unprimed axes S then are parallel and are moving with velocity $-\mathbf{u}$ relative to the frame S'.

Momentarily, the following relations hold for the particle

$$u_x = \mathbf{u};\ a_x = \frac{du}{dt}$$

$$u_y = u_z = 0$$

$$u'_x = u'_y = u'_z = 0$$

Inserting these values in Eqs (2.104), we get

$$a_x = \frac{a'_x}{\gamma^3}$$

$$a_y = \frac{a'_y}{\gamma^2} \quad \text{(i)}$$

$$a_z = \frac{a'_z}{\gamma^2}$$

To an observer in S', the particle of mass m moves subject to the electromagnetic field E' and $\mathbf{B}'$. Since the motion is slow, the observer in S adopts Newtonian mechanics and so $\mathbf{B}'$ has a negligible influence. Hence we can write the equation of motion in the component form as

$$ma'_x = eE'_x$$

$$ma'_y = eE'_y \quad \text{(ii)}$$

$$ma'_z = eE'_z$$

We know from Eqs [6.29(a)–[6.29(c)] that

$$E'_x = E_x$$

$$E'_y = \gamma\left(E_y - \tfrac{u}{c}B_z\right) \quad \text{(iii)}$$

$$E'_z = \gamma\left(E_z + \tfrac{v}{c}B_y\right)$$

Inserting into Eqs (ii), the values of primed quantities in terms of unprimed ones from Eqs (i) and (iii), we obtain

$$m\gamma^3 a_x = eE_x$$

$$m\gamma a_y = e\left(E_y - \tfrac{u}{c}B_z\right) \quad \text{(iv)}$$

$$m\gamma a_z = e\left(E_x + \tfrac{u}{c}B_y\right)$$

The equation of motion is stated in the form originally proposed by Newton,

Rate of change of linear momentum = Force

and so we get

$$\frac{d}{dt}\frac{mu_x}{\sqrt{1-\frac{u^2}{c^2}}} = \frac{m\frac{du_x}{dt}}{\sqrt{1-\frac{u^2}{c^2}}} + \frac{mu_x u\frac{du}{dt}}{c^2\left(1-\frac{u^2}{c^2}\right)^{\frac{3}{2}}}$$

$$= \frac{ma_x}{(1-u^2/c^2)^{3/2}} = m\gamma^3 a_x$$

$$\frac{d}{dt}\frac{mu_y}{\sqrt{1-\frac{u^2}{c^2}}} = \frac{ma_y}{\sqrt{1-\frac{u^2}{c^2}}} = m\gamma a_y \qquad \text{(v)}$$

$$\frac{d}{dt}\frac{mu_z}{\sqrt{1-\frac{u^2}{c^2}}} = \frac{ma_z}{\sqrt{1-\frac{u^2}{c^2}}} = m\gamma a_z$$

Summing up all the quantities of Eqs (v) and making use of Eq. (iv), we get

$$\frac{d}{dt}\left(\frac{m\mathbf{u}}{\sqrt{1-\frac{u^2}{c^2}}}\right) = e\mathbf{E} + \frac{e}{c}(\mathbf{u}\times\mathbf{B})$$

or

$$\mathbf{F} = e\left[\mathbf{E} + \frac{1}{c}(\mathbf{u}\times\mathbf{B})\right]$$

6.3 ELECTROMAGNETIC FIELD TENSOR

In the Lorentz gauge, the electric field and magnetic induction are given in terms of ϕ and $\mathbf{A}$ as

$$\mathbf{E} = -\frac{1}{c}\frac{\partial\mathbf{A}}{\partial t} - \nabla\phi \qquad (6.30)$$

$$\mathbf{B} = \nabla\times\mathbf{A} \qquad (6.31)$$

Let us write the expressions for the components of the vector E in terms of the covariant four-potential $A_\mu = (\phi, -\mathbf{A})$

$$E_x = -\frac{\partial\phi}{\partial x} - \frac{1}{c}\frac{\partial A_x}{\partial t} = \frac{\partial A_1}{\partial x^0} - \frac{\partial A_0}{\partial x^1}$$

$$= \partial_0 A_1 - \partial_1 A_0 \qquad [6.32(a)]$$

where use has been made of the covariant vector operator

$$\partial_\mu \equiv \frac{\partial}{\partial x_\mu} = \left(\frac{\partial}{\partial x^0}, \nabla\right)$$

$$E_y = -\frac{\partial \phi}{\partial y} - \frac{1}{c}\frac{\partial A_y}{\partial t} = \frac{\partial A_2}{\partial x^0} - \frac{\partial A_0}{\partial x^2}$$ [6.32(b)]

$$= \partial_0 A_2 - \partial_2 A_0$$

$$E_z = -\frac{\partial \phi}{\partial z} - \frac{1}{c}\frac{\partial A_z}{\partial t} = \frac{\partial A_3}{\partial x^0} - \frac{\partial A_0}{\partial x^3}$$ [6.32(c)]

$$= \partial_0 A_3 - \partial_3 A_0$$

Further we write the components of vector **B**

$$B_x = \frac{\partial A_z}{\partial y} - \frac{\partial A_y}{\partial z} = \frac{\partial A_2}{\partial x^3} - \frac{\partial A_3}{\partial x^2}$$ [6.33(a)]

$$= \partial_3 A_2 - \partial_2 A_3$$

$$B_y = \frac{\partial A_x}{\partial z} - \frac{\partial A_z}{\partial x} = \frac{\partial A_3}{\partial x^1} - \frac{\partial A_1}{\partial x^3}$$

$$= \partial_1 A_3 - \partial_3 A_1$$ [6.33(b)]

$$B_z = \frac{\partial A_y}{\partial x} - \frac{\partial A_x}{\partial y} = \frac{\partial A_1}{\partial x^2} - \frac{\partial A_2}{\partial x^1}$$

$$= \partial_2 A_1 - \partial_1 A_2$$ [6.33(c)]

It is known from tensor algebra that expressions like $(a_\mu b_\nu - a_\nu b_\mu)$ are the covariant components of an antisymmetric second rank tensor; a_μ and b_ν are arbitrary components of arbitrary vectors. Thus the components of E and **B**, Eqs [6.32(a)]–[6.32(c)] and [6.33(a)]–[6.33(c)], can be expressed as the covariant components of the antisymmetric four-tensor

$$F_{\mu\nu} = \partial_\mu A_\nu - \partial_\nu A_\mu$$

$$= \frac{\partial A_\nu}{\partial x^\mu} - \frac{\partial A_\mu}{\partial x^\nu} \tag{6.34}$$

$F_{\mu\nu}$ is antisymmetric since

$$F_{\nu\mu} = \frac{\partial A_\mu}{\partial x^\nu} - \frac{\partial A_\nu}{\partial x^\mu}$$

$$= -F_{\mu\nu} \tag{6.35}$$

and

$$F_{\mu\mu} = 0 \tag{6.36}$$

$F_{\mu\nu}$ is called the electromagnetic field tensor in the covariant form. The contravariant components of the electromagnetic field tensor $F^{\mu\nu}$ are obtained by raising the indices μ and ν on both the sides of Eq. (6.35) and lowering them on the right-hand side. Thus,

$$F^{\mu\nu} = \frac{\partial A^{\nu}}{dx_{\mu}} - \frac{\partial A^{\mu}}{\partial x_{\nu}} \tag{6.37}$$

Further we evaluate the various components of $F_{\mu\nu}$ from Eq. (6.35)

$$F_{01} = -F_{10} = \frac{\partial A_1}{\partial x^0} - \frac{\partial A_0}{\partial x^1} = -\frac{\partial A_x}{c\partial t} - \frac{\partial \phi}{\partial x} = \mathcal{E}_x$$

$$F_{02} = -F_{20} = \frac{\partial A_2}{\partial x^0} - \frac{\partial A_0}{\partial x^2} = -\frac{\partial A_y}{c\partial t} - \frac{\partial \phi}{\partial y} = \mathcal{E}_y$$

$$F_{03} = -F_{30} = \frac{\partial A_3}{\partial x^0} - \frac{\partial A_0}{\partial x^3} = -\frac{\partial A_z}{c\partial t} - \frac{\partial \phi}{\partial z} = \mathcal{E}_z$$

$$F_{12} = -F_{21} = \frac{\partial A_2}{\partial x^1} - \frac{\partial A_1}{\partial x^2} = -\frac{\partial A_y}{\partial x} + \frac{\partial A_x}{\partial y} = -B_z$$

$$F_{13} = -F_{31} = \frac{\partial A_3}{\partial x^1} - \frac{\partial A_1}{\partial x^3} = -\frac{\partial A_z}{\partial x} + \frac{\partial A_x}{\partial z} = B_y$$

$$F_{23} = -F_{32} = \frac{\partial A_3}{\partial x^2} - \frac{\partial A_2}{\partial x^3} = -\frac{\partial A_z}{\partial y} + \frac{\partial A_y}{\partial z} = -B_x$$

Writing the tensor $F_{\mu\nu}$ in the tabular form, we have

$$F_{\mu\nu} = \begin{vmatrix} 0 & E_x & E_y & E_z \\ -E_x & 0 & -B_z & B_y \\ -E_y & B_z & 0 & -B_x \\ -E_z & -B_y & B_x & 0 \end{vmatrix} \tag{6.38}$$

The contravariant form of the electromagnetic field tensor can be written from Eq. (6.38), noting that raising or lowering of the time index does not change the sign of the component, while raising or lowering the space index reverses the sign of component. Thus,

$$F_{\mu\nu} = \begin{vmatrix} 0 & -E_x & -E_y & -E_z \\ E_x & 0 & -B_z & B_y \\ E_y & B_z & 0 & -B_x \\ E_z & -B_y & B_x & 0 \end{vmatrix} \tag{6.39}$$

The space components of $F_{\mu\nu}$ are the magnetic flux intensities. Since the components of **B** form a three-dimensional antisymmetric tensor of the second rank and it is only an axial vector that can form an antisymmetric tensor of the second rank, thus **B** is an axial vector. The electric field intensities constitute a column or row vector in $F_{\mu\nu}$, thus **E** is a polar vector.

Sometimes we need the dual field-strength tensor $\mathcal{F}^{\mu\nu}$ defined by

$$\mathcal{F}^{\mu\nu} = \frac{1}{2}\varepsilon^{\mu\nu\rho\sigma}\mathcal{F}_{\rho\sigma} \tag{6.40}$$

where $\varepsilon^{\mu\nu\rho\sigma}$ is an antisymmetric unit four-pseudotensor of the fourth rank, defined as follows:

$$\varepsilon^{\mu\nu\rho\sigma} = \begin{cases} 0 \text{ if any two indices are equal} \\ +1 \text{ if } \mu\nu\rho\sigma \text{ is an even permutation of } 0, 1, 2, 3 \\ -1 \text{ if } \mu\nu\rho\sigma \text{ is an odd permutation of } 0, 1, 2, 3 \end{cases}$$

The tensor $\varepsilon^{\mu\nu\rho\sigma}$ has $4! = 24$ components. We get

$$\mathcal{F}^{\mu\nu} = \begin{vmatrix} 0 & -B_x & -B_y & -B_z \\ B_x & 0 & E_z & -E_y \\ B_y & -E_z & 0 & E_x \\ B_z & E_y & -E_x & 0 \end{vmatrix} \tag{6.41}$$

Comparing Eqs (6.39) and (6.41), it is obvious that $\mathcal{F}^{\mu\nu}$ can be obtained from $F^{\mu\nu}$ by replacing $\mathbf{E} \rightarrow \mathbf{B}$ and $\mathbf{B} \rightarrow -\mathbf{E}$.

6.3.1 Electromagnetic Field Invariants

The invariants of an electromagnetic field can be found by contracting a tensor, since contraction over any pair of indices lowers its rank by two. The trace of $F_{\mu\nu}$, $\Sigma F_{\mu\mu}$, the sum of its diagonal elements, is a zero rank tensor, i.e., an invariant. However, since $F_{\mu\nu}$ is an antisymmetric tensor, $\Sigma F_{\mu\mu} = 0$ and there is no linear invariant. Further, we form a quadratic invariant from the product $F_{\mu\nu} F^{\rho\sigma}$, that is a tensor of fourth rank. Let us perform a double contraction by assigning the indices μ and ρ, ν, and σ, equal and summating over these. The double contraction will lower the rank by four and give a tensor of rank zero, i.e., an invariant. $\Sigma F^{\mu\nu} F_{\mu\nu}$ is the sum of the products of the respective elements of the two tensors. Thus,

$$\sum_{\mu,\nu} F_{\mu\nu} F^{\mu\nu} = -2\sum_{k=1}^{3} E_k^2 + 2\sum_{k=1}^{3} B_k^2$$

$$= 2(B^2 - E^2) = \text{invariant} \tag{6.42}$$

This shows that $(B^2 - E^2)$ or $(E^2 - B^2)$ is a Lorentz invariant.
Now, considering the contraction of $F_{\mu\nu}$ and $\mathcal{F}^{\mu\nu}$ and using Eqs (6.38) and (6.41), we get

$$\sum_{\mu,\nu} F_{\mu\nu} \mathcal{F}^{\mu\nu} = -4\,\mathbf{E}\cdot\mathbf{B} \tag{6.43}$$

showing that $\mathbf{E}\cdot\mathbf{B}$ is another Lorentz invariant of electromagnetic field.

Example 6.2 Show that

(i) If **E** and **B** are perpendicular in one Lorentz frame, they are perpendicular in all Lorentz frames.
(ii) If |**E**| > |**B**| in any one Lorentz frame, then |**E**| > |**B**| in any other Lorentz frame and vice versa.
(iii) If the angle between **E** and **B** is acute (or obtuse) in one Lorentz frame, it is acute (or obtuse) in any other Lorentz frame.
(iv) If **E** is perpendicular to **B**, but |**E**| ≠ |**B**| then there is a frame in which the field is either purely electric or purely magnetic.

Solution.

(i) Because of the invariance of **E** and **B** under Lorentz transformation, we infer that if in one reference system $\mathbf{E}\cdot\mathbf{B} = 0$, that is, **E** is perpendicular to **B**, then these are also perpendicular in any other Lorentz frame.
(ii) If |**E**| > |**B**|, i.e., $(E^2 - B^2) > 0$ (or <0) in one Lorentz frame then $(E^2 - B^2) > 0$ (or <0) due to the fact that $E^2 - B^2 =$ invariant.
(iii) Since $\mathbf{E}\cdot\mathbf{B} =$ invariant under Lorentz transformation, so if in any system of reference vectors **E** and **B** make an acute (or obtuse) angle, they will make an acute (or obtuse) angle in every other reference system.
(iv) Now $\mathbf{E}\cdot\mathbf{B} = 0$ is given, then we can always find a reference system in which $\mathbf{E} = 0$ or $\mathbf{B} = 0$ according as $(E^2 - B^2) > 0$ (or < 0), that is, the field is either purely magnetic or purely electric.

6.3.2 Transformation of Electromagnetic Fields

Electromagnetic field tensor $F^{\mu\nu}$ in system S is defined as

$$F^{\mu\nu} = \frac{\partial A^{\nu}}{\partial x_{\mu}} - \frac{\partial A^{\mu}}{\partial x_{\nu}} \tag{6.44}$$

and in system S', it becomes

$$F'^{\mu\nu} = \frac{\partial A'^{\nu}}{\partial x'_{\mu}} - \frac{\partial A'^{\mu}}{\partial x'_{\nu}} \tag{6.45}$$

Let us derive the transformation law for the components of $F^{\mu\nu}$ under Lorentz transformation. Rewriting the Lorentz transformations for x_{μ} and A^{ν}, we have

$$x'_{\mu} = \sum_{\nu=0}^{3} \alpha^{\nu}_{\mu} x_{\nu} \quad (\mu = 0, 1, 2, 3) \tag{6.46}$$

$$A'^{\nu} = \sum_{\mu=0}^{3} \alpha^{\nu}_{\mu} A^{\mu} \quad (\nu = 0, 1, 2, 3) \tag{6.47}$$

$[\alpha^{\mu}_{\nu}]$, the transformation matrix is

$$[\alpha^{\nu}_{\mu}] = \begin{pmatrix} \alpha_0 & \alpha_1 & 0 & 0 \\ \alpha_1 & \alpha_0 & 0 & 0 \\ 0 & 0 & 1 & 0 \\ 0 & 0 & 0 & 1 \end{pmatrix} = \begin{pmatrix} \gamma & -\gamma\beta & 0 & 0 \\ -\gamma\beta & \gamma & 0 & 0 \\ 0 & 0 & 1 & 0 \\ 0 & 0 & 0 & 1 \end{pmatrix} \tag{6.48}$$

Writing the antisymmetric four-tensor explicitly, we have

$$(\mathcal{F}^{\mu\nu}) = \begin{pmatrix} 0 & F^{01} & F^{01} & F^{03} \\ -F^{01} & 0 & F^{12} & F^{13} \\ -F^{02} & -F^{12} & 0 & F^{23} \\ -F^{03} & -F^{13} & -F^{23} & 0 \end{pmatrix} \tag{6.49}$$

as it has only six independent components.
The second rank electromagnetic field tensor $F^{\mu\nu}$ will transform under Lorentz transformation as

$$F'^{\mu\nu} = \sum_{\rho,\sigma=0}^{3} \alpha^{\mu}_{\rho}\, \alpha^{\nu}_{\sigma} F^{\rho\sigma} \tag{6.50}$$

We evaluate the component F'^{01}. Thus,

$$\begin{aligned} F'^{01} &= \sum_{\rho,\sigma=0}^{3} \alpha^{0}_{\rho}\alpha^{1}_{\sigma} F^{\rho\sigma} \\ &= \sum_{\rho=0}^{3} \alpha^{0}_{\rho} \sum_{\sigma=0}^{3} \alpha^{1}_{\sigma} F^{\rho\sigma} \\ &= \alpha^{0}_{0} \sum_{\sigma=0}^{3} \alpha^{1}_{\sigma} F^{0\sigma} + \alpha^{0}_{1} \sum_{\sigma=0}^{3} \alpha^{1}_{\sigma} F^{1\sigma} \\ &\quad + \alpha^{0}_{2} \sum_{\sigma=0}^{1} \alpha^{1}_{\sigma} F^{2\sigma} + \alpha^{0}_{3} \sum_{\sigma=0}^{3} \alpha^{1}_{\sigma} F^{3\sigma} \end{aligned}$$

From Eq. (6.49), we see that $\alpha^{0}_{2} = \alpha^{0}_{3} = 0$ so it becomes

$$F'^{01} = \alpha^{0}_{0}\left(\alpha^{1}_{0} F^{00} + \alpha^{1}_{1} F^{01}\right) + \alpha^{0}_{1}\left(\alpha^{1}_{0} F^{10} + \alpha^{1}_{1} F^{11}\right) \tag{6.51}$$

From Eq. (6.49), it is obvious that

$$F^{00} = F^{11} = 0;\; F^{10} = -F^{01} \tag{6.52}$$

and substituting α_0 for α^{0}_{0} and α^{1}_{1}, also α_1 for α^{1}_{0} and α^{0}_{1}; we have

$$\begin{aligned} F'^{01} &= (\alpha^{2}_{0} - \alpha^{2}_{1}) F^{01} \\ &= F^{01} \end{aligned} \tag{6.53}$$

since $(\alpha^{2}_{0} - \alpha^{2}_{1}) = 1$.

Proceeding analogously, one can calculate the transformation formulae for all the other five components. These are

$$F'^{01} = F^{01} \tag{6.54}$$

$$F'^{02} = \alpha_0 F^{02} + \alpha_1 F^{12} \tag{6.55}$$

$$F'^{03} = \alpha_0 F^{03} + \alpha_1 F^{13} \tag{6.56}$$

$$F'^{12} = \alpha_0 F^{12} + \alpha_1 F^{02} \tag{6.57}$$

$$F'^{13} = \alpha_0 F^{13} + \alpha_1 F^{03} \tag{6.58}$$

$$F'^{23} = F^{23} \tag{6.59}$$

Substituting the values of different elements of the matrix ($F^{\mu\nu}$), Eq. (6.39), we have

$$E'_x = E_x \qquad [6.60(a)]$$

$$E'_y = \gamma(E_y - \beta B_z) \qquad [6.60(b)]$$

$$E'_z = \gamma(E_z - \beta B_y) \qquad [6.60(c)]$$

and

$$B'_x = B_x \qquad [6.61(a)]$$

$$B'_y = \gamma(E_y + \beta E_z) \qquad [6.61(b)]$$

$$B'_z = \gamma(E_z + \beta E_y) \qquad [6.61(c)]$$

The inverse transformations can be written by the usual prescription of changing the sign of β and primed quantities into unprimed and vice versa.

One can generalize these transformations for an arbitrary direction of relative velocity. Resolving the field in directions parallel and perpendicular to the direction of relative translation, we have the transformation equations for the general case as

$$\mathbf{E}'_\perp = \gamma\left(\mathbf{E}_\perp + \boldsymbol{\beta}\times\mathbf{B}_\perp\right) \tag{6.62}$$

$$\mathbf{E}'_\parallel = \mathbf{E}_\parallel \tag{6.63}$$

$$\mathbf{B}'_\perp = \gamma\left(\mathbf{B}_\perp + \boldsymbol{\beta}\times\mathbf{E}_\perp\right) \tag{6.64}$$

$$\mathbf{B}'_\parallel = \mathbf{B}_\parallel \tag{6.65}$$

The set of Eqs (6.60) and (6.61) for non-relativistic case, i.e. $v \ll c$, are

$$E'_x = E_x \qquad [6.66(a)]$$

$$E'_y = E_y - \beta B_z \qquad [6.66(b)]$$

$$E'_z = E_z + \beta B_y \qquad [6.66(c)]$$

and

$$B'_x = B_x \qquad [6.67(a)]$$

$$B'_y = E_y + \beta E_z \qquad [6.67(b)]$$

$$B'_z = E_z - \beta E_y \qquad [6.67(c)]$$

Expressing these in vector form, we get

$$\mathbf{E}' = \mathbf{E} + \boldsymbol{\beta} \times \mathbf{B} \qquad (6.68)$$

$$\mathbf{B}' = \mathbf{B} - \boldsymbol{\beta} \times \mathbf{E} \qquad (6.69)$$

It is evident from these transformation equations that **E** and **B** have no independent existence as separate entities. The fundamental complex is the electromagnetic field tensor $F^{\mu\nu}$ and the values of **E** and **B** are totally relative to the motion of the observer.

6.4 COVARIANT FORM OF MAXWELL'S EQUATIONS

The fundamental form of Maxwell's equations is

$$\nabla \times \mathbf{E} + \frac{1}{c}\frac{\partial \mathbf{B}}{\partial t} = 0 \qquad (6.70)$$

$$\nabla \cdot \mathbf{B} = 0 \qquad (6.71)$$

$$\nabla \times \mathbf{B} - \frac{1}{c}\frac{\partial \mathbf{E}}{\partial t} = \frac{4\pi}{c}\mathbf{J} \qquad (6.72)$$

$$\nabla \cdot \mathbf{E} = 4\pi\rho \qquad (6.73)$$

where the symbols have their usual meanings. When these equations are expressed in four-dimensional form, the electric and magnetic field intensities turn out to be components of electromagnetic field tensor.

Let us first consider the inhomogeneous set of Eqs (6.72) and (6.73)

$$\nabla \times \mathbf{B} - \frac{1}{c}\frac{\partial \mathbf{E}}{\partial t} = \frac{4\pi}{c}\mathbf{J} \qquad (6.74)$$

$$\nabla \cdot \mathbf{E} = 4\pi\rho \qquad (6.75)$$

Expressed in the form of components, above equations can be written as

$$0 + \frac{\partial E_x}{\partial x^1} + \frac{\partial E_y}{\partial x^2} + \frac{\partial E_z}{\partial x^3} = \frac{4\pi}{c} j^0 \qquad [6.76(a)]$$

$$-\frac{\partial E_x}{\partial x^0} + 0 + \frac{\partial B_z}{\partial x^2} - \frac{\partial B_y}{\partial x^3} = \frac{4\pi}{c} j^1 \qquad [6.76(b)]$$

$$-\frac{\partial E_y}{\partial x^0} - \frac{\partial B_z}{\partial x^1} + 0 + \frac{\partial B_x}{\partial x^3} = \frac{4\pi}{c} j^2 \qquad [6.76(c)]$$

$$-\frac{\partial E_x}{\partial x^0} + \frac{\partial B_y}{\partial x^1} - \frac{\partial B_x}{\partial x^2} + 0 = \frac{4\pi}{c} j^3 \qquad [6.76(d)]$$

Introducing on the left-hand side, a set of variables defined as

$$F^{00} = 0, F^{01} = -E_x, F^{02} = -E_y, F^{03} = -E_z \quad [6.77(a)]$$

$$F^{10} = E_x, F^{11} = 0, F^{12} = -B_z, F^{13} = B_y \quad [6.77(b)]$$

$$F^{20} = E_y, F^{21} = B_z, F^{22} = 0, F^{23} = -B_x \quad [6.77(c)]$$

$$F^{30} = E_z, F^{31} = -B_y, F^{32} = B_x, F^{23} = 0 \quad [6.77(d)]$$

and $j^\mu = (j^0, j^1, j^2, j^3)$ in the right-hand side, Eqs [6.76(a)–[6.76(d)] become

$$\sum_{\nu=0}^{3} \frac{\partial F^{\mu\nu}}{\partial x^\nu} = -\frac{4\pi}{c} j^\mu \quad (\mu = 0, 1, 2, 3) \quad (6.78)$$

where

$$(F^{\mu\nu}) = \begin{vmatrix} 0 & -E_x & -E_y & -E_z \\ E_x & 0 & -B_z & B_y \\ E_y & B_z & 0 & -B_x \\ E_z & -B_y & B_x & 0 \end{vmatrix} \quad (6.79)$$

is the contravariant antisymmetric tensor of the second rank.

Let us now consider the set of homogeneous equations

$$\nabla \times \mathbf{E} + \frac{1}{c}\frac{\partial \mathbf{B}}{\partial t} = 0 \quad [6.80(a)]$$

$$\nabla \cdot \mathbf{B} = 0 \quad [6.80(b)]$$

Writing Eq. [6.41] in the component form, we have

$$\frac{\partial B_x}{\partial x} + \frac{\partial B_y}{\partial y} + \frac{\partial B_z}{\partial z} = 0 \quad (6.81)$$

Expressing it in terms of $\mathcal{F}^{\mu\nu}$, Eq. (6.41), we get

$$\partial_1 \mathcal{F}^{10} + \partial_2 \mathcal{F}^{20} + \partial_3 \mathcal{F}^{30} = 0 \quad (6.82)$$

or

$$\sum_{\mu=0}^{3} \partial_\mu \mathcal{F}^{\mu 0} = 0 \quad (6.83)$$

Furthermore, the x-component of Eq. [6.80(a)] is

$$\frac{\partial E_z}{\partial y} - \frac{\partial E_y}{\partial z} + \frac{1}{c}\frac{\partial B_x}{\partial t} = 0 \quad (6.84)$$

When written in terms of field tensors, it becomes

$$\partial_2 F^{30} + \partial_3 F^{02} + \partial_0 F^{32} = 0 \tag{6.85}$$

and

$$\sum_{\mu=0}^{3} \partial_\mu \mathcal{F}^{\mu 1} = 0 \tag{6.86}$$

One can generalize Eqs (6.82) and (6.85) immediately as

$$\sum_{\mu=0}^{3} \partial_\mu \mathcal{F}^{\mu\nu} = 0 \tag{6.87}$$

It represents the homogeneous set of Eqs [6.80(a)] and [6.80(b)]. The indices 1, 2, 3 appear in a cyclic order in Eq. (6.82) but the indices 0, 2, 3 in Eq. (6.85) do not appear in a cyclic manner as the third term does not follow the cyclic order. However, this shortcoming can be taken care of, if we use the contravariant differentiation operator $\partial^\mu \equiv \frac{\partial}{\partial x_\mu} = \left(\frac{\partial}{\partial x_0}, -\nabla\right)$ instead of the covariant operator $\partial_\mu \equiv \frac{\partial}{\partial x^\mu} = \left(\frac{\partial}{\partial x^0}, \nabla\right)$. Thus, Eq. (6.85) can be written as

$$\partial^2 F^{30} + \partial^3 F^{02} + \partial^0 F^{23} = 0 \tag{6.88}$$

As is evident, the indices 0, 2, 3 appear in cyclic manner in all the three terms. However, Eq. (6.82) is not affected as all the terms get multiplied by -1 and it becomes

$$\partial^1 F^{23} + \partial^2 F^{31} + \partial^3 F^{12} = 0 \tag{6.89}$$

Equations (6.88) and (6.89) can be generalized as

$$\partial^\mu F^{\nu\rho} + \partial^\nu F^{\rho\mu} + \partial^\rho F^{\mu\nu} = 0 \tag{6.90}$$

where the indices μ, ν, ρ have the range (0, 1, 2, 3).

Let us show that Eq. (6.90) represents the y- and z-components of Eq. [6.80(a)]. We write Eq. (6.90) for $\mu = 0$, $\nu = 1$, and $\rho = 2$.

$$\frac{\partial F^{12}}{\partial x^0} - \frac{\partial F^{20}}{\partial x^1} - \frac{\partial F^{01}}{\partial x^2} = 0$$

$$-\frac{\partial B_z}{c\partial t} - \frac{\partial E_y}{\partial x} + \frac{\partial E_x}{\partial y} = 0$$

or

$$\frac{\partial E_y}{\partial x} - \frac{\partial F_x}{\partial y} + \frac{\partial B_z}{c\partial t} = 0 \tag{6.91}$$

This is the z-component of equation $\nabla \times \mathbf{E} + \frac{1}{c}\frac{\partial \mathbf{B}}{\partial t} = 0$. Analogously, we get

$$-\frac{\partial F^{01}}{\partial x_3} + \frac{\partial F^{13}}{\partial x_0} - \frac{\partial F^{30}}{\partial x_1} = 0$$

$$\frac{\partial E_x}{\partial z} + \frac{\partial B_y}{c\partial t} - \frac{\partial E_z}{\partial x} = 0$$

or

$$\frac{\partial E_x}{\partial z} - \frac{\partial E_z}{\partial x} + \frac{\partial B_y}{c\partial t} = 0 \tag{6.92}$$

This represents the y-component of equation

$$\nabla \times \mathbf{E} + \frac{1}{c}\frac{\partial \mathbf{B}}{\partial t} = 0.$$

To summarize, Maxwell's equations in the covariant form are

$$\sum_{\nu=0}^{3} \frac{\partial F^{\mu\nu}}{\partial x^{\nu}} = -\frac{4\pi}{c} j^{\mu} \quad (\mu = 0, 1, 2, 3) \tag{6.93(a)}$$

$$\sum_{\mu=0}^{3} \partial_{\mu} \mathcal{F}^{\mu\nu} = 0 \tag{6.93(b)}$$

or

$$\sum_{\nu=0}^{3} \frac{\partial F^{\mu\nu}}{\partial x^{\nu}} = -\frac{4\pi}{c} j^{\mu} \tag{6.94(a)}$$

$$\partial^{\mu} F^{\nu\rho} + \partial^{\nu} F^{\rho\mu} + \partial^{\rho} F^{\mu\nu} = 0 \tag{6.94(b)}$$

6.5 LAGRANGIAN AND HAMILTONIAN FOR A CHARGED PARTICLE IN AN ELECTROMAGNETIC FIELD

$$\nabla \times \mathbf{E} + \frac{1}{c}\frac{\partial \mathbf{B}}{\partial t} = 0 \tag{6.95(a)}$$

$$\nabla \times \mathbf{B} - \frac{1}{c}\frac{\partial \mathbf{E}}{\partial t} = \frac{4\pi}{c}\mathbf{J} \tag{6.95(b)}$$

$$\nabla \cdot \mathbf{E} = 4\pi\rho \tag{6.95(c)}$$

$$\nabla \cdot \mathbf{B} = 0 \tag{6.95(d)}$$

Let a particle of mass m and charge e lie in an electromagnetic field characterized by the vector potential $\mathbf{A}$ and scalar potential ϕ. The force on a charge e in an electromagnetic field is given by Lorentz force

$$\mathbf{F} = e\left[\mathbf{E} + \frac{1}{c}(\mathbf{u} \times \mathbf{B})\right] \tag{6.96}$$

From $\nabla\cdot\mathbf{B} = \mathbf{0}$, it follows that

$$\mathbf{B} = \nabla \times \mathbf{A} \tag{6.97}$$

From Eq. [6.95(a)], we get

$$\nabla \times \mathbf{E} + \frac{1}{c}\frac{\partial}{\partial t}(\nabla \times \mathbf{A}) = \nabla \times \left(\mathbf{E} + \frac{1}{c}\frac{\partial \mathbf{A}}{\partial t}\right) = 0$$

so

$$\mathbf{E} + \frac{1}{c}\frac{\partial \mathbf{A}}{\partial t} = -\nabla\phi$$

or

$$\mathbf{E} = -\nabla\phi - \frac{1}{c}\frac{\partial \mathbf{A}}{\partial t} \tag{6.98}$$

In terms of potentials ϕ and $\mathbf{A}$, Lorentz force equation becomes

$$\mathbf{F} = e\left[-\nabla\phi - \frac{1}{c}\frac{\partial \mathbf{A}}{\partial t} + \frac{1}{c}(\mathbf{u} \times \nabla \times \mathbf{A})\right] \tag{6.99}$$

The terms of Eq. (6.99) can be written in a more convenient form. Let us consider the x-component.

$$(\nabla\phi)_x = \frac{\partial\phi}{\partial x} \tag{6.100}$$

$$(\mathbf{u} \times \nabla \times \mathbf{A})_x = u_y\left(\frac{\partial A_y}{\partial x} - \frac{\partial A_x}{\partial y}\right) - u_z\left(\frac{\partial A_x}{\partial z} - \frac{\partial A_z}{\partial x}\right) \tag{6.101}$$

$$= u_y\frac{\partial A_y}{\partial x} + u_z\frac{\partial A_z}{\partial x} + u_x\frac{\partial A_x}{\partial x} - u_y\frac{\partial A_x}{\partial_y} - u_z\frac{\partial A_x}{\partial z} - u_x\frac{\partial A_x}{\partial x} \tag{6.102}$$

Now

$$\frac{dA_x}{dt} = \frac{\partial A_x}{\partial t} + \left(u_x\frac{\partial A_x}{\partial x} + u_y\frac{\partial A_x}{\partial y} + u_z\frac{\partial A_x}{\partial z}\right) \tag{6.103}$$

Thus the x-component of $(\mathbf{u} \times \nabla \times \mathbf{A})$ is given by

$$(\mathbf{u} \times \nabla \times \mathbf{A})_x = \frac{\partial(\mathbf{u}\cdot\mathbf{A})}{\partial x} - \frac{dA_x}{dt} + \frac{\partial A_x}{\partial t} \tag{6.104}$$

Rewriting Eq. (6.99)

$$F_x = e\left[-\frac{\partial}{\partial x}\left(\phi - \frac{1}{c}(\mathbf{u}\cdot\mathbf{A})\right) - \frac{1}{c}\frac{d}{dt}\left(\frac{\partial}{\partial u_x}\right)(\mathbf{A}\cdot\mathbf{u})\right]$$

This expression is equivalent to

$$F_x = -\frac{\partial U}{\partial x} + \frac{d}{dt}\frac{\partial U}{\partial u_x} \tag{6.105}$$

where

$$U = e\phi - \frac{e}{c}\mathbf{A}\cdot\mathbf{u} \tag{6.106}$$

and U is a generalized potential.
The Lagrangian for a charged particle in an electromagnetic field is

$$\mathcal{L} = -mc^2\sqrt{1 - u^2/c^2} + \frac{e}{c}\mathbf{A}\cdot\mathbf{u} - e\phi \tag{6.107}$$

where $-mc^2(1 - u^2/c^2)^{1/2}$ is the Lagrangian for the free particle, Eq. (4.4). The last two terms describe the interaction of the charge with the field.
The generalized momentum of the particle is defined as

$$\mathbf{P} = \frac{\partial \mathcal{L}}{\partial \mathbf{u}} = \frac{m\mathbf{u}}{\sqrt{1 - u^2/c^2}} + \frac{e}{c}\mathbf{A}$$

$$= \mathbf{p} + \frac{e}{c}\mathbf{A} \tag{6.108}$$

where $\mathbf{p} = m\mathbf{u}$ is the ordinary momentum.
The Hamiltonian for a particle in an electromagnetic field is defined as

$$\mathcal{H} = \mathbf{u}\cdot\frac{\partial \mathcal{L}}{\partial \mathbf{u}} - \mathcal{L} \tag{6.109}$$

$$= \mathbf{u}\cdot\frac{m\mathbf{u}}{\sqrt{1 - u^2/c^2}} + \frac{e}{c}\mathbf{u}\cdot\mathbf{A} + mc^2\sqrt{1 - u^2/c^2} - \frac{e}{c}\mathbf{u}\cdot\mathbf{A} + e\phi$$

$$= \frac{mc^2}{\sqrt{1 - u^2/c^2}} + e\phi \tag{6.110}$$

It is more conventional to express the Hamiltonian in terms of generalized momentum **P**, rather than velocity. From Eq. (6.110), we get

$$\left(\frac{\mathcal{H} - e\phi}{c}\right)^2 = \frac{m^2c^2}{(1 - u^2/c^2)} = \frac{m^2u^2 + m^2(c^2 - u^2)}{(1 - u^2/c^2)}$$

$$= \frac{m^2u^2}{(1 - u^2/c^2)} + m^2c^2 \tag{6.111}$$

Comparing it with Eq. (6.108), we obtain

$$\left(\frac{\mathcal{H} - e\phi}{c}\right)^2 = m^2c^2 + \left(\mathbf{P} - \frac{e}{c}\mathbf{A}\right)^2$$

or

$$\mathcal{H} = \sqrt{m^2c^4 + c^2\left(\mathbf{P} - \frac{e}{c}\mathbf{A}\right)^2} + e\phi \tag{6.112}$$

Example 6.3 Prove that Hamiltonian for a charged particle in an electromagnetic field is given by

$$\mathcal{H} = \sqrt{m^2c^4 + c^2\left(\mathbf{P} - \tfrac{e}{c}\mathbf{A}\right)^2} + e\phi.$$

Solution. Hamilton's canonical equations are

$$\frac{\partial \mathcal{H}}{\partial p_x} = \dot{x}$$

$$\frac{\partial \mathcal{H}}{\partial x} = -\dot{p}_x$$

where a dot denotes differentiation with respect to time *t*. These equations should give the correct equations of motion from the above Hamiltonian.

$$\frac{\partial \mathcal{H}}{\partial p_x} = \frac{c\left(P_x - \tfrac{e}{c}A_x\right)}{\sqrt{m^2c^2 + \left(\mathbf{P} - \tfrac{e}{c}\mathbf{A}\right)^2}}$$

$$= \frac{c\left(P_x - \tfrac{e}{c}A_x\right)\sqrt{1 - \frac{u^2}{c^2}}}{mc} = \frac{cmu_x}{\sqrt{1 - \frac{u^2}{c^2}}} \times \frac{\sqrt{1 - \frac{u^2}{c^2}}}{mc}$$

$$= \dot{x}, \text{ etc.}$$

The second equation

$$-\dot{p}_x = \frac{\partial \mathcal{H}}{\partial x}$$

gives

$$-\frac{d}{dt}\frac{mu_x}{\sqrt{1 - \frac{u^2}{c^2}}} - \frac{e}{c}\left(\frac{\partial A_x}{\partial x}u_x + \frac{\partial A_x}{\partial y}u_y + \frac{\partial A_x}{\partial z}u_z + \frac{\partial A_x}{\partial t}\right)$$

$$= e\frac{\partial \phi}{\partial x} - \frac{e}{c}\left(u_x\frac{\partial A_x}{\partial x} + u_y\frac{\partial A_y}{\partial x} + u_z\frac{\partial A_z}{\partial x}\right)$$

that is

$$\frac{d}{dt}\left(\frac{mu_x}{\sqrt{1 - \frac{u^2}{c^2}}}\right) = -e\frac{\partial \phi}{\partial x} - \frac{e}{c}\frac{\partial A_x}{\partial t} + \frac{e}{c}\left[u_y\left(\frac{\partial A_y}{\partial x} - \frac{\partial A_x}{\partial y}\right) - u_z\left(\frac{\partial A_x}{\partial z} - \frac{\partial A_z}{\partial x}\right)\right]$$

$$= eE_x + \frac{e}{c}(\mathbf{u} \times \mathbf{B})_x$$

Similarly,

$$\frac{d}{dt}\frac{mu_y}{\sqrt{1-\frac{u^2}{c^2}}} = eE_y + \frac{e}{c}(\mathbf{u}\times\mathbf{B})_y$$

$$\frac{d}{dt}\frac{mu_z}{\sqrt{1-u^2/c^2}} = eE_z + \frac{e}{c}(\mathbf{u}\times\mathbf{B})_z$$

Summing the above three equations, we get

$$\frac{d}{dt}\left(\frac{m\mathbf{u}}{\sqrt{1-u^2/c^2}}\right) = \mathbf{F} = e\mathbf{E} + \frac{e}{c}(\mathbf{u}\times\mathbf{B})$$

which is the Lorentz force equation. Thus, the given Hamiltonian is the correct one.

Example 6.4 How will you express the scalar potential of a uniform electric field in terms of the field intensity **E** and the vector potential of a uniform magnetic field in terms of magnetic induction **B**.

Solution. A field is said to be uniform if the field intensities are the same at all points in space. For a uniform electric field, the potential $\phi = -\mathbf{E}\cdot\mathbf{r}$, where **E** is constant.

Since
$$\mathbf{E} = -\nabla\phi = \nabla(\mathbf{E}\cdot\mathbf{r})$$
$$= (\mathbf{E}\cdot\nabla) + \mathbf{E}\times(\nabla\times\mathbf{r})$$
$$= \mathbf{E}$$

The vector potential for a uniform magnetic field $\mathbf{A} = 1/2\ \mathbf{B}\times\mathbf{r}$, where **B** is constant.

We have
$$\mathbf{B} = \nabla\times\mathbf{A}$$
$$= \nabla\times(1/2\ \mathbf{B}\times\mathbf{r})$$
$$= 1/2\nabla\times(\mathbf{B}\times\mathbf{r})$$
$$= \frac{1}{2}[\mathbf{B}(\nabla\cdot\mathbf{r}) - (\mathbf{B}\cdot\nabla)\mathbf{r}]$$
$$= \mathbf{B}$$

6.6 EQUATION OF MOTION OF A CHARGE IN AN ELECTROMAGNETIC FIELD

6.6.1 Three-dimensional Vector Form

The equation of motion will be the Lagrange equations

$$\frac{d}{dt}\left(\frac{\partial\mathcal{L}}{\partial\mathbf{u}}\right) = \frac{\partial\mathcal{L}}{\partial\mathbf{r}} \tag{6.113}$$

where

$$\mathcal{L} = -mc^2\sqrt{1 - u^2/c^2} + \frac{e}{c}\,\mathbf{A}\cdot\mathbf{u} - e\phi$$

Now
$$\frac{\partial \mathcal{L}}{\partial \mathbf{r}} = \nabla\mathcal{L} = \frac{e}{c}\,\nabla(\mathbf{A}\cdot\mathbf{u}) - e\nabla\phi \tag{6.114}$$

and
$$\frac{\partial \mathcal{L}}{\partial \mathbf{u}} = \mathbf{P} = \mathbf{p} + \frac{e}{c}\,\mathbf{A} \tag{6.115}$$

Making use of result from vector formula

$$\nabla(\mathbf{A}\cdot\mathbf{B}) = (\mathbf{A}\cdot\nabla)\mathbf{B} + (\mathbf{B}\cdot\nabla)\mathbf{A} + \mathbf{B}\times\nabla\times\mathbf{A} + \mathbf{A}\times\nabla\times\mathbf{B}$$

So
$$\nabla(\mathbf{A}\cdot\mathbf{u}) = (\mathbf{u}\cdot\nabla)\mathbf{A} + \mathbf{u}\times\nabla\times\mathbf{A} \tag{6.116}$$

Thus from Eq. (6.113), we have

$$\frac{d}{dt}\left(\mathbf{p} + \frac{e}{c}\,\mathbf{A}\right) = \frac{e}{c}\,(\mathbf{u}\cdot\nabla)\mathbf{A} + \frac{e}{c}\,\mathbf{u}\times\nabla\times\mathbf{A} - e\nabla\phi \tag{6.117}$$

But
$$\frac{d\mathbf{A}}{dt} = \frac{\partial\mathbf{A}}{\partial t} + (\mathbf{u}\cdot\nabla)\mathbf{A} \tag{6.118}$$

Therefore we get

$$\frac{d\mathbf{p}}{dt} = -\frac{e}{c}\frac{\partial\mathbf{A}}{\partial t} + \frac{e}{c}\,(\mathbf{u}\times\nabla\times\mathbf{A}) - e\nabla\phi \tag{6.119}$$

Using the results we get

$$\mathbf{E} = -\frac{1}{c}\frac{\partial\mathbf{A}}{\partial t} - \nabla\phi$$

and
$$\mathbf{B} = \nabla\times\mathbf{A}$$

Finally we set

$$\frac{d\mathbf{p}}{dt} = e\mathbf{E} + \frac{e}{c}\,\mathbf{u}\times\mathbf{B} \tag{6.120}$$

The expression on the right-hand side is called the Lorentz force. It consists of two parts, the first term represents the force exerted by the electric field and is along **E**, the second is proportional to the velocity **u** of the charge and is at right angles to the plane containing **u** and **B**. As the force which the magnetic field exerts on the charge is always at right angles to its velocity, the magnetic field does no work on the charge moving under its influence. On the contrary, the electric field does work and the rate of change of kinetic energy is derived as follows:

$$\frac{d\mathbf{E}_{kin}}{dt} = \frac{d}{dt}\left(\frac{mc^2}{\sqrt{1 - u^2/c^2}}\right) \tag{6.121}$$

Now $$d\mathcal{E}_{kin} = \mathbf{u}\cdot d\mathbf{p}$$

Therefore $$\frac{d\mathcal{E}_{kin}}{dt} = \mathbf{u}\cdot\frac{d\mathbf{p}}{dt}$$

$$= \mathbf{u}\cdot\left[e\mathbf{E} + \frac{e}{c}\,\mathbf{u}\times\mathbf{B}\right]$$

$$= e\mathbf{E}\cdot\mathbf{u} \tag{6.122}$$

Let us examine the effect of time reversal on Eq. (6.120)

$$\frac{d\mathbf{p}}{dt} = e\left[\mathbf{E} + \frac{1}{c}\,\mathbf{u}\times\mathbf{B}\right] \tag{6.123}$$

In classical mechanics, when t is replaced by $-t$, the equations of motion retain the same form, i.e., they are invariant under this operation of time reversal. However, in the case of electromagnetic field, the time reversal will hold provided when changing t by $-t$, we also change $\mathbf{B}$ to $-\mathbf{B}$ and $\mathbf{E}$ to $\mathbf{E}$.

Since $$\mathbf{B} = \nabla\times\mathbf{A}$$

$$-\mathbf{B} = -\nabla\times\mathbf{A} = \nabla\times(-\mathbf{A}) \tag{6.124}$$

Thus the vector potential changes sign whereas the scalar potential does not.

6.6.2 Four-dimensional Form

The action function for a charge in an electromagnetic field is

$$S = \int_1^2\left(-mcds - \frac{e}{c}\sum_{\mu=0}^{3} A_\mu dx^\mu\right) \tag{6.125}$$

m is the mass of a particle, e its charge, and A_μ is the four-potential of the field. The first term denotes the action for a free particle and the second term describes the interaction of the particle with the electromagnetic field. It may be remarked that the field being determinate, its action will not vary and as such is not included. However, this is strictly tenable only if the charge of the particle is so small that it does not modify the field. It is assumed that this condition is fulfilled.

The principle of least action states that

$$\delta S = \delta\int_1^2\left(-mc\sqrt{dx_\mu dx^\mu} - \frac{e}{c}A_\mu dx^\mu\right) = 0 \tag{6.126}$$

where we substituted $ds^2 = \sum dx_\mu dx^\mu$. Thus,

$$\delta S = \int_1^2\left(-mc\sum\frac{dx_\mu d\delta x^\mu}{ds} - \frac{e}{c}\sum A_\mu d\delta x^\mu - \frac{e}{c}\sum \delta A_\mu dx^\mu\right) = 0 \tag{6.127}$$

Here we substitute $\delta dx^\mu = d\delta x^\mu$.

Integrating by parts the first two terms in the integrand, we have

$$\delta S = -m\sum \frac{dx_\mu}{d\tau}\,\delta x^\mu\Big|_1^2 + \int_1^2 m\sum \delta x^\mu \frac{dx_\mu}{d\tau}\,d\tau - \frac{e}{c}\sum A_\mu \delta x^\mu\Big|_1^2$$

$$+\int_1^2 \frac{e}{c}\sum \delta x^\mu dA_\mu - \int_1^2 \frac{e}{c}\sum \delta A_\mu dx^\mu = 0 \tag{6.128}$$

where τ is the proper time of the particle and u_μ is its four-velocity. The first and third terms vanish because at the ends of the trajectory $\delta x^\mu = -\delta x_\mu = 0$. Thus,

$$\delta S = \int_1^2 \left(m\sum_\mu \delta x^\mu \frac{du_\mu}{d\tau}\,d\tau + \frac{e}{c}\sum_\mu \delta x^\mu dA_\mu - \frac{e}{c}\sum_\nu \delta A_\nu dx^\nu \right) = 0 \tag{6.129}$$

The dummy index in the last term has been denoted by ν, so that we could put it in terms of $F_{\mu\nu}$. Substituting

$$dA_\mu = \sum_\nu \frac{\partial A_\mu}{\partial x^\nu}\,dx^\nu = \sum_\nu \frac{\partial A_\mu}{\partial x^\nu}\,u^\nu d\tau \tag{6.130}$$

$$\delta A_\nu = \sum_\mu \frac{\partial A_\nu}{\partial x^\mu}\delta x^\mu \tag{6.131}$$

$$dx^\nu = u^\nu\,d\tau \tag{6.132}$$

we get

$$\delta S = -\int_1^2 \left(m\sum_\mu \delta x^\mu \frac{du_\mu}{d\tau} + \frac{e}{c}\sum_\mu \delta x^\mu \sum_\nu \frac{\partial A_\mu}{\partial x^\nu}\,u^\nu d\tau - \frac{e}{c}\sum_\nu u^\nu d\tau \sum_\mu \frac{\partial A_\nu}{\partial x^\mu}\delta x^\mu \right) = 0 \tag{6.133}$$

Rewriting it, we get

$$\delta S = \int_1^2 \sum_\mu \left(m\frac{du_\mu}{d\tau} + \frac{e}{c}\sum_\nu \frac{\partial A_\mu}{\partial x^\nu}\,u^\nu - \frac{e}{c}\sum_\nu \frac{\partial A_\nu}{\partial x^\mu}\,u^\nu \right) \delta x^\mu d\tau = 0 \tag{6.134}$$

As δx^μ is arbitrary, the above result will hold only when

$$m\frac{du_\mu}{d\tau} = \frac{e}{c}\sum \left(\frac{\partial A_\mu}{\partial x^\nu} - \frac{\partial A_\nu}{\partial x^\mu} \right) u^\nu$$

or

$$m\frac{du_\mu}{d\tau} = \frac{e}{c}\sum_\nu F_{\mu\nu}u^\nu \tag{6.135}$$

This is the equation of motion of a charge in electromagnetic field in four-dimensional form obtained through the calculus of variations.

The above equation may be expressed in terms of Minkowski four-vector f_μ, defined by the relation

$$f_\mu = \frac{1}{c}\sum_\nu F_{\mu\nu} j^\nu \tag{6.136}$$

Since

$$\sum_{\nu=0}^{3} \frac{\partial F_{\mu\nu}}{\partial x^\nu} = -\frac{4\pi}{c} j^\mu \quad (\mu = 0, 1, 2, 3) \tag{6.137}$$

therefore

$$f_\mu = -\frac{1}{4\pi}\sum_{\nu,\sigma=0}^{3} F_{\mu\nu}\frac{\partial F^{\sigma\nu}}{\partial x^\sigma} \tag{6.138}$$

Example 6.5 Show that the orbit of a particle of charge e, rest mass m in a uniform electrostatic field $\mathcal{E}$ along the x-axis is

$$x = \frac{U_0}{eE}\cosh\frac{eE_y}{cp_0}$$

where $U_0 = \sqrt{(mc^2)^2 + c^2 p_0^2}$ and p_0 is the initial momentum in the y-direction. The momentum in the x-direction at $y = 0$ has been taken equal to zero.

Solution. Consider the motion of a particle of charge e in a uniform electric field $\mathbf{E}$, which is directed along the x-axis. The equation of motion is

$$\frac{d\mathbf{p}}{dt} = e\mathbf{E}$$

which in the component form becomes

$$\frac{dp_x}{dt} = eE; \quad \frac{dp_y}{dt} = 0$$

which gives $\quad p_x = eEt;\ p_y = p_0$

$t = 0$ corresponds to the moment when $p_x = eEt$; p_0 is the momentum of the particle in the y-direction at $t = 0$.

The energy of the particle, $U = \sqrt{m^2c^4 + c^2p_0^2 + (ceEt)^2}$. The velocity $\mathbf{u}$ of the particle is

$$\mathbf{u} = \frac{\mathbf{p}c^2}{U}$$

Thus for $u_x = \dot{x}$, we have

$$\frac{dx}{dt} = \frac{c^2p_0}{U} = \frac{c^2eEt}{\sqrt{U_0^2 + (ceEt)^2}}$$

where $U_0 = \sqrt{m^2c^4 + c^2p_0^2}$ is the energy at $t = 0$.

Integrating, we find that

$$x = \frac{1}{eE}\sqrt{U_0^2 + (ceEt)^2}$$

setting the constant of integration equal to zero.

Further,

$$\frac{dy}{dt} = \frac{p_y c^2}{U} = \frac{p_0 c^2}{\sqrt{U_0^2 + (ceEt)^2}}$$

$$y = \frac{p_0 c}{eE} \sinh^{-1}\left(\frac{ceEt}{U_0}\right)$$

For obtaining the equation of trajectory, we have to eliminate t from equations for x and y, so that

$$x = \frac{1}{eE}\sqrt{U_0^2 + (ceEt)^2}$$

$$= \frac{1}{eE}\sqrt{U_0^2 + \left(\frac{ceE}{ceE} U_0 \sinh \frac{yeE}{p_0 c}\right)^2}$$

$$= \frac{U_0}{eE}\sqrt{1 + \sinh^2 \frac{yeE}{p_0 c}}$$

$$= \frac{U_0}{eE} \cosh \frac{yeE}{p_0 c}$$

Thus in a uniform electric field, a charge moves along a centenary curve.

Example 6.6 *Thomson velocity selector:* For selecting electrons of one particular speed, which could be calculated from the fields, J. J. Thomson employed the arrangement of electric and magnetic fields at right angles. Calculate the constant velocity.

Solution. If the electrons have the proper velocity **u** at right angles to both the fields, the magnetic force balances the electrostatic force and the electrons therefore travel in a straight line with constant speed.

The electrostatic force is given by

$$\mathbf{F}_e = e\mathbf{E}$$

and the magnetic force by

$$\mathbf{F}_m = \frac{e}{c}\mathbf{u} \times \mathbf{B}$$

For proper u,

$$\mathbf{F}_e + \mathbf{F}_m = 0$$

$$e\left[\mathbf{E} + \frac{1}{c}(\mathbf{u} \times \mathbf{B})\right] = 0$$

or

$$e\left[\mathbf{E} + \frac{1}{c}(\mathbf{u} \times \mathbf{B})\right] \times \mathbf{B} = 0$$

Since **u** and **B** are at right angles, we have

$$(\mathbf{u} \times \mathbf{B}) \times \mathbf{B} = -B^2\mathbf{u}$$

Hence,

$$\mathbf{E} \times \mathbf{B} - \frac{B^2\mathbf{u}}{c} = 0$$

or

$$\mathbf{u} = \frac{c(\mathbf{E} \times \mathbf{B})}{B^2}$$

6.7 MAXWELL'S EQUATIONS BY VARIATIONAL TECHNIQUE

For finding the field equations from the calculus of variations, we assume that the motion of charges is given and we vary only the potentials that play the role of coordinates of the system. The expression for the action of the system, field and particles, will consist of three terms:

$$S = S_f + S_m + S_{mf} \tag{6.139}$$

where S_f is the part of action that depends on the field without any charges, S_m is the part that depends on the free particles, and the term S_{mf} accounts for action that depends on the interaction between the particles and field.

Let us consider these parts one by one. S_f is the action of the electromagnetic field. We know that the electromagnetic field obeys the superposition principle. Thus the equations for the field must be linear differential equations. The equations of a field are obtained by variation of action. As in variation of action, the power of the integrand gets lowered by one degree, the action for the field should contain terms quadratic in the field. Two simple quadratic invariants suggest immediately, i.e., $\sum_{\mu=0}^{3} A_\mu A^\mu$ and $\sum_{\mu,\nu=0}^{3} F_{\mu\nu} F^{\mu\nu}$. Since the four-potential is always indeterminate within the gauge transformation, the first invariant is not suitable. Thus the only invariant of the field that recommends itself is

$$S_f = \alpha \int \sum_{\mu,\nu} F^{\mu\nu} F_{\mu\nu} dV^* \tag{6.140}$$

where α is a constant and $dV^* = cdVdt$. In order to obtain the correct equations of the field (in Gaussian system), one has to substitute $\alpha = -\frac{1}{16\pi c}$. Therefore,

$$S_f = -\frac{1}{16\pi c} \int \sum_{\mu,\nu} F_{\mu\nu} F^{\mu\nu} dV^* \tag{6.141}$$

The remaining two terms are given for all the particles, Eq. (6.125)

$$S_m = -\sum_{a=1}^{n} m_a c \int ds_a \tag{6.142}$$

$$S_{mf} = -\sum_{a=1}^{n} \frac{e_a}{c} \int \sum_{\mu=0}^{3} A_{\mu a} dx_a^\mu \tag{6.143}$$

where a is the number of the particle, n is the total number of particles, and $A_{\mu a}$ is the potential of the field at the point of the four-space where ath particle is. Rewriting Eq. (6.143)

$$S_{mf} = -\sum_{a=1}^{n} \frac{e_a}{c} \sum \sum_{\mu=0}^{3} A_{\mu a} \frac{dx_a^{\mu}}{dt} dt \tag{6.144}$$

Introducing a charge density ρ for the distribution of point charges e_a as

$$\rho = \sum_{a=1}^{n} e_a \delta(\mathbf{r} - \mathbf{r}_a) \tag{6.145}$$

we get

$$S_{mf} = -\frac{1}{c} \int \rho dV \int \sum_{\mu=0}^{3} A_{\mu} \frac{dx^{\mu}}{dt} dt \tag{6.146}$$

Substituting $\rho \left(\frac{dx^{\mu}}{dt} \right) = j^{\mu}$, one gets

$$S_{mf} = -\frac{1}{c} \int \sum_{\mu} A_{\mu} j^{\mu} dV dt$$

$$= -\frac{1}{c^2} \int \sum_{\mu} A_{\mu} j^{\mu} dV^* \tag{6.147}$$

where $dV^* = dx^0 dx^1 dx^2 dx^3 = cdVdt$.

Finally, the expression for action for the system, field and particles, is

$$S = -\sum_a m_a c \int ds_a - \frac{1}{c^2} \int \sum_{\mu} A_{\mu} j^{\mu} dV^* - \frac{1}{16\pi c} \int \sum_{\mu,\nu} F_{\mu\nu} F^{\mu\nu} dV^* \tag{6.148}$$

For obtaining the equations of the field, we assume that the motion of the charges is specified; as such the term S_m may be neglected. Therefore we calculate the variation of the expression

$$S = S_f + S_{mf}$$

$$= -\frac{1}{16\pi c} \int \sum_{\mu,\nu} F_{\mu\nu} F^{\mu\nu} dV^* - \frac{1}{c^2} \int \sum_{\mu} A_{\mu} j^{\mu} dV^* \tag{6.149}$$

As the motion of charges is pre-specified, j^{μ} does not vary. Thus, the variation of action is given by

$$\delta S = -\frac{1}{16\pi c} \int \delta \sum_{\mu,\nu} F_{\mu\nu} F^{\mu\nu} dV^* - \frac{1}{c^2} \int \sum_{\mu} j^{\mu} \delta A_{\mu} dV^* \tag{6.150}$$

Now

$$\delta \sum_{\mu,\nu} F_{\mu\nu} F^{\mu\nu} = \sum_{\mu,\nu} F_{\mu\nu} \delta F^{\mu\nu} + \sum_{\mu,\nu} F^{\mu\nu} \delta F_{\mu\nu}$$

$$= \sum_{\mu,\nu} F^{\mu\nu} \delta F_{\mu\nu} + \sum_{\mu,\nu} F^{\mu\nu} \delta F_{\mu\nu}$$

$$= 2\sum_{\mu,\nu} F^{\mu\nu}\delta F_{\mu\nu}$$

$$= 2\sum_{\mu,\nu} F^{\mu\nu}\delta\left(\frac{\partial A_\nu}{\partial x^\mu} - \frac{\partial A_\mu}{\partial x^\nu}\right)$$

$$= 2\sum_{\mu,\nu} F^{\mu\nu}\,\delta\frac{\partial A_\nu}{\partial x^\mu} - 2\sum_{\mu,\nu} F^{\mu,\nu}\frac{\partial A_\mu}{\partial x^\nu} \tag{6.151}$$

Substituting $F^{\mu\nu} = -F^{\nu\mu}$ and then interchanging the indices μ and ν in the first term on the right-hand side, we get

$$\delta\sum_{\mu,\nu} F_{\mu\nu}F^{\mu\nu} = -2\sum_{\mu,\nu} F^{\mu\nu}\,\delta\frac{\partial A_\mu}{\partial x^\nu} - 2\sum_{\mu,\nu} F^{\mu\nu}\,\delta\frac{\partial A_\mu}{\partial x^\nu}$$

$$= -4\sum_{\mu,\nu} F^{\mu\nu}\delta\frac{\partial A_\mu}{\partial x^\nu} \tag{6.152}$$

Further, interchanging the sequence of variation and differentiation we have

$$\delta\sum_{\mu,\nu} F_{\mu\nu}F^{\mu\nu} = -4\sum_{\mu,\nu} F^{\mu\nu}\frac{\partial}{\partial x^\nu}\delta A_\mu$$

$$= -4\sum_{\mu,\nu}\frac{\partial}{\partial x^\nu}(F^{\mu\nu}\,\delta A_\mu) + 4\sum_{\mu,\nu}\delta A_\mu\frac{\partial F^{\mu\nu}}{\partial x^\nu} \tag{6.153}$$

The variation of action becomes

$$\delta S = \frac{1}{4\pi c}\int\sum_{\mu,\nu}\frac{\partial}{\partial x^\nu}(F^{\mu\nu}\,\delta A^\mu)\,dV^* - \frac{1}{4\pi c}\int\sum_{\mu,\nu}\delta A_\mu\frac{\partial F^{\mu\nu}}{\partial x^\nu}dV^* - \frac{1}{c^2}\int\sum_\mu j^\mu\,\delta A_\mu dV^* \tag{6.154}$$

By Gauss's theorem, the first integral is transformed into a surface integral

$$\delta S = \int\sum_{\mu,\nu}\frac{\partial}{\partial x^\nu}(F^{\mu\nu}\,\delta A_\mu)\,dV^* = \int\sum_{\mu,\nu}F^{\mu\nu}\,\delta A_\mu df_\nu \tag{6.155}$$

As $\delta A_\mu = 0$ at the boundary of the four-volume being considered, it vanishes. Thus,

$$\delta S = -\frac{1}{c}\int\sum_\mu\left(\frac{1}{4\pi}\sum_\nu\frac{\partial F^{\mu\nu}}{\partial x^\nu} + \frac{1}{c}j^\mu\right)\delta A_\mu dV^* \tag{6.156}$$

The variation δA_μ being arbitrary, δS will be zero only if the integrand is zero. Therefore,

$$\sum_\nu\frac{\partial F^{\mu\nu}}{\partial x^\nu} = -\frac{4\pi}{c}j^\mu \quad (\mu = 0, 1, 2, 3) \tag{6.157}$$

Let us express these four equations in three-dimensional form. For $\mu = 0$, it gives

$$\frac{\partial F^{00}}{\partial x^0} + \frac{\partial F^{01}}{\partial x^1} + \frac{\partial F^{02}}{\partial x^2} + \frac{\partial F^{03}}{\partial x^3} = -\frac{4\pi}{c}j^0$$

Substituting the values for the components of $F^{\mu\nu}$, we get

$$-\frac{\partial E_x}{\partial x}-\frac{\partial E_y}{\partial y}-\frac{\partial E_z}{\partial z}=-4\pi\rho$$

or

$$\nabla\cdot\mathbf{E}=4\pi\rho \tag{6.158}$$

For $\mu=1$, we obtain

$$\frac{\partial F^{10}}{\partial x^0}+\frac{\partial F^{11}}{\partial x^1}+\frac{\partial F^{12}}{\partial x^2}+\frac{\partial F^{13}}{\partial x^3}=-\frac{4\pi}{c}j^1$$

$$\frac{\partial E_x}{c\partial t}-\frac{\partial B_z}{\partial y}+\frac{\partial B_y}{\partial z}=-\frac{4\pi}{c}j_x$$

or

$$\frac{\partial B_z}{\partial y}-\frac{\partial B_y}{\partial z}-\frac{1}{c}\frac{\partial E_x}{\partial t}=\frac{4\pi}{c}j_x \tag{6.159}$$

This with the two equations for $\mu=2, 3$ can be written as

$$\nabla\times\mathbf{B}-\frac{1}{c}\frac{\partial\mathbf{E}}{\partial t}=\frac{4\pi}{c}\mathbf{J} \tag{6.160}$$

Equations (6.158) and (6.160) constitute a pair of inhomogeneous Maxwell's equations.

6.8 ELECTROMAGNETIC ENERGY–MOMENTUM TENSOR

Let us derive an expression for the energy–momentum tensor of the electromagnetic field. We make use of the expression for Minkowski force, Eq. (6.136), given by

$$f_\mu=-\frac{1}{c}\sum_\nu F_{\mu\nu}j^\nu \tag{6.161}$$

Substituting for j^ν from Eq. (6.137), one obtains

$$f_\mu=\frac{1}{4\pi}\sum_{\nu,\,\sigma=0}^{3}F_{\mu\nu}\frac{\partial F^{\nu\sigma}}{\partial x^\sigma} \tag{6.162}$$

$$=\frac{1}{4\pi}\sum_{\nu,\,\sigma=0}^{3}\left[\frac{\partial}{\partial x^\sigma}\left(F_{\mu\nu}F^{\nu\sigma}\right)-F^{\nu\sigma}\frac{\partial F_{\mu\nu}}{\partial x^\sigma}\right] \tag{6.163}$$

Since ν, σ are dummy indices, we interchange them in the second term in the bracket and get

$$F^{\nu\sigma}\frac{\partial F_{\mu\nu}}{\partial x^\sigma}=F^{\sigma\nu}\frac{\partial F_{\mu\sigma}}{\partial x^\nu} \tag{6.164}$$

Making use of the antisymmetric properties of tensors $F^{\sigma\nu}$ and $F_{\sigma\nu}$, we obtain

$$F^{\nu\sigma}\frac{\partial F_{\mu\nu}}{\partial x^{\sigma}} = F^{\sigma\nu}\frac{\partial F_{\mu\sigma}}{\partial x^{\nu}}$$

$$= \frac{1}{2}F^{\nu\sigma}\left[\frac{\partial F_{\mu\nu}}{\partial x^{\sigma}} + \frac{\partial F_{\sigma\mu}}{\partial x^{\nu}}\right] \tag{6.165}$$

Since $\frac{\partial F_{\mu\nu}}{\partial x^{\sigma}} + \frac{\partial F_{\sigma\mu}}{\partial x^{\nu}} + \frac{\partial F_{\nu\sigma}}{\partial x^{\mu}} = 0$ we write Eq. (6.165), as follows

$$F^{\nu\sigma}\frac{\partial F_{\mu\nu}}{\partial x^{\sigma}} = -\frac{1}{2}F^{\nu\sigma}\frac{\partial F_{\nu\sigma}}{\partial x^{\mu}}$$

$$= -\frac{1}{4}\frac{\partial}{\partial x^{\mu}}\left(F^{\nu\sigma}F_{\nu\sigma}\right) \tag{6.166}$$

Therefore, Eq. (6.163) becomes

$$f_{\mu} = \frac{1}{4\pi}\sum_{\nu,\,\sigma=0}^{3}\left[\frac{\partial}{\partial x^{\sigma}}\left(-F_{\mu\nu}F^{\sigma\nu}\right) + \frac{1}{4}\,\delta_{\mu}^{\sigma}\frac{\partial}{\partial x^{\sigma}}\left(F^{\nu\rho}F_{\nu\rho}\right)\right] \tag{6.167}$$

where we used the relations: $F^{\nu\sigma} = -F^{\sigma\nu}$ and $\frac{\partial}{\partial x^{\mu}} = \delta_{\mu}^{\sigma}\frac{\partial}{\partial x^{\sigma}}$

Defining

$$S_{\mu}^{\sigma} = -\frac{1}{4\pi}\left[\sum_{\nu=0}^{3}F_{\mu\nu}F^{\sigma\nu} - \frac{1}{4}\,\delta_{\mu}^{\sigma}\sum_{\lambda,\,\rho=0}^{3}F_{\lambda\rho}F^{\lambda\rho}\right] \tag{6.168}$$

we write Eq. (6.167) as

$$f_{\mu} = \sum_{\sigma}\frac{\partial S_{\mu}^{\sigma}}{\partial x^{\sigma}} \tag{6.169}$$

where S_{μ}^{σ} is called the energy momentum tensor. In order to express it in the covariant form $S^{\mu\sigma}$, we raise the index μ and $\delta_{\mu}^{\sigma} \to g^{\mu\sigma}$, thus

$$S^{\mu\sigma} = -\frac{1}{4\pi}\sum_{\nu}F_{\ \nu}^{\mu}F^{\sigma\nu} + \frac{1}{16\pi}\,g^{\mu\nu}\sum_{\lambda,\,\rho=0}^{3}F_{\lambda\rho}F^{\lambda\rho} \tag{6.170}$$

It can be shown that the contravariant and covariant components of a unit tensor are customarily designated by $g^{\mu\nu}$ and $g_{\mu\nu}$ and these are

$$(g^{\mu\nu}) = (g_{\mu\nu}) = \begin{vmatrix} 1 & 0 & 0 & 0 \\ 0 & -1 & 0 & 0 \\ 0 & 0 & -1 & 0 \\ 0 & 0 & 0 & -1 \end{vmatrix} \tag{6.171}$$

Now,

$$S^{\sigma\mu} = -\frac{1}{4\pi}\sum_{\nu} F^{\sigma}_{\ \nu} F^{\mu\nu} + \frac{1}{16\pi} g^{\sigma\mu} \sum_{\lambda,\rho=0}^{3} F_{\lambda\rho} F^{\lambda\rho} \tag{6.172}$$

If an index is raised in one factor and the same index is simultaneously lowered in the second factor, the expression remains unaltered. Thus, comparing Eqs (6.171) and (6.172), we see that energy–momentum tensor is symmetric, i.e., $S^{\mu\sigma} = S^{\sigma\mu}$.
Further we show that the tensor $S^{\mu\sigma}$ is traceless, i.e., $\sum_{\mu} S^{\mu\mu} = 0$
According to Eq. (6.168),

$$\sum_{\mu} S^{\mu}_{\ \mu} = \sum_{\mu}\left[-\frac{1}{4\pi}\sum_{\nu} F_{\mu\nu} F^{\mu\nu}\right] + \sum_{\mu}\left[\frac{1}{16\pi}\,\delta^{\mu}_{\ \mu}\sum_{\lambda,\rho=0}^{3} F_{\lambda\rho} F^{\lambda\rho}\right]$$

Since $\sum_{\mu} \delta^{\mu}_{\ \mu} = 4$, we get

$$\sum_{\mu} S^{\mu}_{\ \mu} = -\frac{1}{4\pi}\sum_{\mu,\nu} F_{\mu\nu} F^{\mu\nu} + \frac{1}{4\pi}\sum_{\lambda,\rho} F_{\lambda\rho} F^{\lambda\rho} \tag{6.173}$$

$$= 0$$

showing thereby that the tensor $S^{\mu\sigma}$ is traceless.

Let us express the tensor $S^{\mu\sigma}$ in the component form in terms of E and $\mathbf{B}$. For this purpose, we introduce the components of electromagnetic field tensor $F^{\mu\nu}$, Eq. (6.39), remembering that lowering the time index does not change the sign whereas lowering the space index reverses the sign of the component. In addition, we see from Eq. (6.42) that $\sum_{\mu,\nu} F_{\mu\nu}F^{\mu\nu} = 2(B^2 - E^2)$, which is an invariant of the electromagnetic field.

Let us calculate the term S^{00} from Eq. (6.170). Thus,

$$S^{00} = -\frac{1}{4\pi}\sum_{\nu} F^{0}_{\ \nu} F^{0\nu} + \frac{1}{16\pi} 2(B^2 - \mathcal{E}^2)$$

$$= -\frac{1}{4\pi}(-F^2) + \frac{1}{8\pi}(B^2 - \mathcal{E}^2)$$

$$S^{00} = \frac{1}{8\pi}(\mathcal{E}^2 + B^2) = \mathcal{E} \tag{6.174}$$

Next we calculate the term S^{01}. Thus,

$$S^{01} = -\frac{1}{4\pi}\sum_{\nu} F^{0}_{\ \nu} F^{1\nu}$$

$$= \frac{1}{4\pi}(\mathbf{E}_y B_z - \mathbf{E}_z B_y)$$

$$= \frac{1}{4\pi}(\mathbf{E}\times\mathbf{B})_x = \frac{1}{c}(\mathbf{S})_x \tag{6.175}$$

where $\mathbf{S} = \frac{1}{4\pi}(\mathbf{E}\times\mathbf{B})$ is the Poynting vector.

Similarly we can show that

$$S^{02} = \frac{S_y}{c} \tag{6.176}$$

$$S^{03} = \frac{S_z}{c} \tag{6.177}$$

Furthermore, we calculate the term S^{12}. Since $g^{12} = 0$, we obtain

$$\begin{aligned} S^{12} &= -\frac{1}{4\pi}\sum_{\nu} F^{2}_{\nu} F^{1\nu} \\ &= -\frac{1}{4\pi}(E_x E_y + B_x B_y) \\ &= S_{xy} \end{aligned} \tag{6.178}$$

where S_{xy} is the component of Maxwell stress tensor. Continuing in a similar manner, one can calculate other components of tensor $S^{\mu\sigma}$ and show that these agree with the relevant components of Maxwell stress tensor. Therefore,

$$S^{ij} = \frac{1}{4\pi}\left[\frac{1}{2}(E^2 + B^2)\delta_{ij} - E_i E_j - B_i B_j\right] \tag{6.179}$$

Tensor $S^{\mu\sigma}$ is finally expressed as

$$(S^{\mu\sigma}) = \begin{pmatrix} \mathcal{E} & S/c \\ S/c & (S_{ij}) \end{pmatrix} \tag{6.180}$$

where the space portion of the tensor $(S^{\mu\sigma})$ is the Maxwell stress tensor (S_{ij}).

6.8.1 Conservation Laws of Vacuum Electrodynamics

The four divergence of energy momentum tensor $S^{\mu\sigma}$ is given by Eq. (6.169) as

$$f_\mu = \sum_\sigma \frac{\partial S^{\sigma}_{\mu}}{\partial x^\sigma} \tag{6.181}$$

or

$$f^\mu = \sum_\sigma \frac{\partial S^{\mu\sigma}}{\partial x^\sigma} \tag{6.182}$$

Let us calculate f^0. Thus,

$$f^0 = \sum_{\sigma=0}^{3} \frac{\partial S^{0\sigma}}{\partial x^\sigma}$$

$$= \frac{\partial S^{00}}{\partial x^0} + \sum_{i=0}^{3} \frac{\partial S^{0i}}{\partial x^i}$$

Substituting S^{00} from Eq. (6.174) and S^{ij} from Eq. (6.179), we get

$$S^{00} = \mathcal{E} \tag{6.183}$$

$$\sum_{i=0}^{3} \frac{\partial S^{0i}}{\partial x^i} = \frac{1}{4\pi} \nabla \cdot (\mathbf{E} \times \mathbf{B}) = \frac{1}{c} \nabla \cdot \mathbf{S} \tag{6.184}$$

$$f^0 = -\frac{1}{c} \mathbf{E} \cdot \mathbf{J} \tag{6.185}$$

Therefore,

$$-\frac{1}{c} \mathbf{E} \cdot \mathbf{J} = \frac{1}{c} \frac{\partial \mathcal{E}}{\partial t} + \frac{1}{c} \nabla \cdot \mathbf{S} \tag{6.186}$$

which is the equation of energy flow and **S** is the Poynting vector.

The law of momentum conservation for the system consisting of electromagnetic field and particles may be obtained from the space portion of Eq. (6.181). Thus,

$$\begin{aligned} f^i &= \sum_{\sigma=0}^{3} \frac{\partial S^{i\sigma}}{\partial x^\sigma} \\ &= \sum_{j=1}^{3} \frac{\partial S^{ij}}{\partial x^j} + \frac{\partial S^{i0}}{\partial x^0} \end{aligned} \tag{6.187}$$

The first term on the right-hand side represents the three-dimensional divergence of the Maxwell stress tensor and

$$S^{i0} = \frac{1}{c} S \tag{6.188}$$

So that, Eq. (6.187) may be rewritten as

$$\mathbf{f} = \nabla \cdot (\mathbf{S}) - \frac{1}{c^2} \frac{\partial \mathbf{S}}{\partial t} \tag{6.189}$$

If V is the volume in which all the charged particles are enclosed and if the field outside the volume V is zero, then integrating Eq. (6.189) over the volume, we obtain

$$\int_V \left(\mathbf{f} + \frac{1}{c^2} \frac{\partial \mathbf{S}}{\partial t} \right) dV = \int_V (\nabla \cdot \mathbf{S}) dV \tag{6.190}$$

The volume integral of the force density **f** gives the total force which is the time derivative of mechanical momentum **p**, thus

$$\int_V \mathbf{f} dV = \frac{d\mathbf{p}}{dt} \tag{6.191}$$

$$\frac{1}{c^2} \mathbf{S} = \mathbf{g} = \frac{1}{4\pi c} (\mathbf{E} \times \mathbf{B}) \tag{6.192}$$

and

$$\int_V \mathbf{g} dV = \mathbf{G} \tag{6.193}$$

where **g** is the momentum density of the electromagnetic field and **G**, the total momentum. Rewriting Eq. (6.190) in view of Eqs (6.191)–(6.193), we get

$$\frac{d}{dt}(\mathbf{p}+\mathbf{G}) = \int_V \nabla\cdot\mathbf{S}dV$$

$$= \int(\mathbf{S})\cdot\mathbf{n}da \tag{6.194}$$

If the field outside the volume V vanishes, then it also vanishes at the boundary encompassing the volume, hence

$$\frac{d}{dt}(\mathbf{p}+\mathbf{G}) = 0 \tag{6.195}$$

which expresses the law of momentum conservation for the combined system of the particles and field.

However, if the field does not vanish on the boundary of the volume V, then $\int_S(\mathbf{S})\cdot\mathbf{n}da$ is the outward flow of momentum per unit area of surface S surrounding V.

6.9 FIELD OF A CHARGE IN UNIFORM MOTION

A point charge q is moving with velocity **u** and we call the laboratory system, the system S, whereas the system of reference moving with the charge, the system S'. Since the coordinate axes can be chosen arbitrarily, let us choose the x-axis of the system S along the direction of motion of the point charge, so that the x'-axis moves relative to x-axis and the axes y- and z- are parallel to the y' and z'-axes.

If x', y', z' are the coordinates of the charge in the system S', then the scalar potential $\phi' = \frac{q}{R'}$ and the vector potential $\mathbf{A}' = 0$, where $R'^2 = x'^2 + y'^2 + z'^2$.

Let the coordinates of the charge q in the S system be $x = ut$, $y = z = 0$. The scalar and vector potentials in the S system can be written from the transformation equations for the components of a four-vector, Eq. (6.17), viz.,

$$\phi = \gamma\left(\phi' + \frac{u}{c}A'_x\right) \tag{6.196(a)}$$

$$A_x = \gamma\left(A'_x + \frac{u}{c}\,\phi'\right) \tag{6.196(b)}$$

$$A_y = A'_y \tag{6.196(c)}$$

$$A_z = A'_z \tag{6.196(d)}$$

where $\gamma = (1 - u^2/c^2)^{-1/2}$ is the Lorentz contraction factor.
Therefore, the scalar potential in S is given by

$$\phi = \gamma\phi'$$

$$= \frac{\gamma q}{R'} \tag{6.197}$$

Expressing R' in terms of the coordinates x, y, z in the S system, we have

$$x' = \gamma(x - ut)$$

$$y' = y;\ z' = z$$

Thus,

$$R'^2 = \gamma^2(x - ut)^2 + y^2 + z^2 \tag{6.198}$$

Substituting this in Eq. (6.197), we find

$$\phi = \frac{q}{R^*}$$

where

$$R^{*2} = (x - ut)^2 + \gamma^{-2}(y^2 + z^2) \tag{6.199}$$

Furthermore, the vector potential **A** in the system S is

$$\mathbf{A} = \frac{q\mathbf{u}}{cR^*} \tag{6.200}$$

Let us compute the fields **E**, **B** and **E**′, **B**′ in the systems S and S', respectively,

$$\mathbf{B}' = 0 \tag{6.201}$$

$$\mathbf{E}' = \frac{q\mathbf{R}'}{R'^3} \tag{6.202}$$

Employing the field component transformation equations, we evaluate the field components E_x, E_y, E_z from the values of **E**′ and **B**′.

Rewriting the field component transformation equations as

$$E_x = E'_x \qquad [6.203(a)]$$

$$E_y = \gamma\left(E'_y + \tfrac{u}{c}B'_z\right) \qquad [6.203(b)]$$

$$E_z = \gamma\left(E'_z - \tfrac{u}{c}B'_y\right) \qquad [6.203(c)]$$

we get

$$E_x = E'_x = \frac{qx'}{R'^3} \qquad [6.204(a)]$$

$$E_y = \gamma E'_y = \frac{\gamma q y'}{R'^3} \qquad [6.204(b)]$$

$$E_z = \frac{\gamma q z'}{R'^3} \qquad [6.204(c)]$$

Writing vectorially for **E**, we have

$$\mathbf{E} = \left(1 - \frac{u^2}{c^2}\right)\frac{q\mathbf{R}}{R^{*3}} \tag{6.205}$$

where **R** is the radius vector from the source point q to the field point (x, y, z), (Figure 6.1).

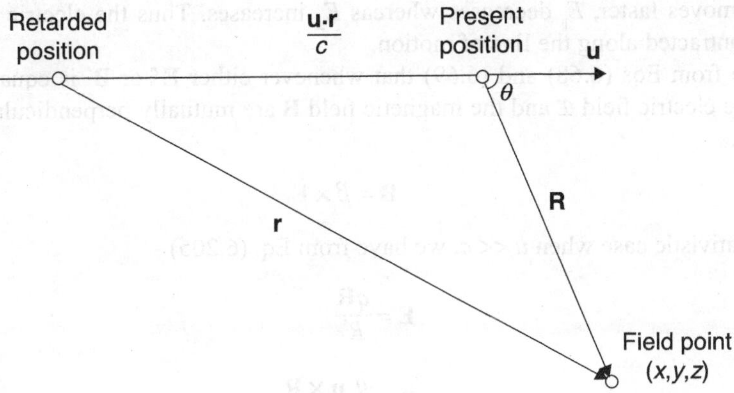

Figure 6.1 *Charge q in uniform motion.*

From Fig. 6.1, we see that

$$y^2 + z^2 = R^2 \sin^2\theta$$

and so R^* becomes

$$\begin{aligned} R^{*2} &= (x - ut)^2 + \gamma^{-2}(y^2 + z^2) \\ &= (x - ut)^2 + \left(1 - \frac{u^2}{c^2}\right)(y^2 + z^2) \\ &= (x - ut)^2 + y^2 + z^2 - \frac{u^2}{c^2}(y^2 + z^2) \\ &= R^2\left(1 - \frac{u^2}{c^2}\right)\sin^2\theta \end{aligned} \tag{6.206}$$

Therefore, from Eq. (6.205), we get

$$\mathbf{E} = \frac{q\mathbf{R}\left(1 - \frac{u^2}{c^2}\right)}{R^3\left(1 - \frac{u^2}{c^2}\sin^2\theta\right)^{3/2}} \tag{6.207}$$

The value of field **E** depends upon θ for a fixed distance R from the source and increases as θ increases from 0 to $\pi/2$ or decreases as θ decreases from π and $\pi/2$.

Now, $$E_{\parallel}(\theta = 0, \pi) = \frac{q}{R^2}\left(1 - \frac{u^2}{c^2}\right) \tag{6.208(a)}$$

This is the smallest value of the field.

and $$E_{\perp}\left(\theta = \frac{\pi}{2}\right) = \frac{q}{R^2}\frac{1}{\sqrt{1 - \frac{u^2}{c^2}}} \tag{6.208(b)}$$

which is the largest value of the field.

As the charge moves faster, $E_{\parallel}$ decreases whereas $E_{\perp}$ increases. Thus the electric field of a moving charge looks contracted along the line of motion.

One can see from Eqs (6.68) and (6.69) that whenever either $\mathbf{E}'$ or $\mathbf{B}'$ is equal to zero in the S' system, then the electric field $\mathcal{E}$ and the magnetic field $\mathbf{B}$ are mutually perpendicular to each other in system S. Thus,

$$\mathbf{B} = \beta \times \mathbf{E} \tag{6.209}$$

For the non-relativistic case when $u << c$, we have from Eq. (6.205)

$$\mathbf{E} = \frac{q\mathbf{R}}{R^3} \tag{6.210}$$

$$\mathbf{B} = \frac{q}{c}\frac{\mathbf{u} \times \mathbf{R}}{R^3} \tag{6.211}$$

Thus, calculations of $\mathbf{E}$ and $\mathbf{B}$ from $\mathbf{E}'$ and $\mathbf{B}'$ are an illustration of the application of Lorentz transformation to the fields of a static charge.

6.10 RETARDED POTENTIALS. LIENARD-WIECHERT POTENTIALS

Let us derive equations determining the potentials for an arbitrary electromagnetic field. Rewriting Maxwell's equations, Eqs (6.94),

$$\sum_{\nu=0}^{3} \frac{\partial F_{\mu\nu}}{\partial x^{\nu}} = -\frac{4\pi}{c} j^{\mu} \tag{6.212}$$

But,
$$F_{\mu\nu} = \frac{\partial A_{\nu}}{\partial x^{\mu}} - \frac{\partial A_{\mu}}{\partial x^{\nu}}$$

Hence,
$$\frac{\partial^2 A_{\nu}}{\partial x^{\mu} \partial x^{\nu}} - \frac{\partial^2 A_{\mu}}{\partial x^{\nu 2}} = -\frac{4\pi}{c} j^{\mu} \tag{6.213}$$

The potentials $\mathbf{A}$ and ϕ are subject to the Lorentz condition

$$\nabla \cdot \mathbf{A} + \frac{1}{c}\frac{\partial \phi}{\partial t} = 0$$

or in four-dimensional form

$$\frac{\partial A_{\mu}}{\partial x^{\mu}} = 0 \tag{6.214}$$

Therefore, Eq. (6.213) becomes

$$\frac{\partial^2 A_{\mu}}{\partial x^{\nu 2}} = -\frac{4\pi}{c} j^{\mu} \tag{6.215}$$

which in three-dimensional form reduces to the set of two equations for **A** and ϕ as

$$\nabla^2 \mathbf{A} - \frac{1}{c^2}\frac{\partial^2 \mathbf{A}}{\partial t^2} = -\frac{4\pi}{c}\mathbf{J} \tag{6.216}$$

$$\nabla^2 \phi - \frac{1}{c^2}\frac{\partial^2 \phi}{\partial t^2} = -4\pi\rho \tag{6.217}$$

which determine the potentials of an arbitrary electromagnetic field. The general solution of these non-homogeneous linear equations is represented as the sum of the complementary function (the solution of these equations without the right-hand side), and the particular integral of these equations with the right-hand side. Let us first determine the particular integral. For this purpose, divide the whole space into infinitely small volume elements and find the field produced by the charges in one of these elements. The total field will be the sum of the fields produced by all such elements. If the charge located in one element of volume under discussion is $de(t)$, then the charge density is $\rho = de(t)\delta(R)$, where R is the distance from the origin. So Eq. (6.217) becomes

$$\nabla^2 \phi - \frac{1}{c^2}\frac{\partial^2 \phi}{\partial t^2} = -4\pi\, de(t)\delta(R) \tag{6.218}$$

Now, $\delta(R) = 0$ except when $R = 0$. Hence everywhere, except at the origin, the equation is

$$\nabla^2 \phi - \frac{1}{c^2}\frac{\partial^2 \phi}{\partial t^2} = 0 \tag{6.219}$$

or in spherical coordinates

$$\frac{1}{R^2}\frac{\partial}{\partial R}\left(R^2\frac{\partial \phi}{\partial R}\right) - \frac{1}{c^2}\frac{\partial^2 \phi}{\partial t^2} = 0 \tag{6.220}$$

Substituting $\phi = \dfrac{\chi(R, t)}{R}$ we get

$$\frac{d^2\chi}{dt^2} - \frac{1}{c^2}\frac{\partial^2 \chi}{\partial t^2} = 0 \tag{6.221}$$

which is a plane wave equation, whose solutions are

$$\chi = f_1\left(t - \frac{R}{c}\right) + f_2\left(t + \frac{R}{c}\right) \tag{6.222}$$

where f_1 is the plane wave moving along positive x-axis and f_2, the wave moving along negative x-axis. Neglecting f_2 on physical grounds, we get

$$\phi = \frac{\chi\left(t - \frac{R}{c}\right)}{R} \tag{6.223}$$

But near the origin $R = 0$, Eq. (6.218) goes over to

$$\nabla^2\phi = -4\pi de(t) \tag{6.224}$$

which gives $\phi = de(t)/R$, the Coulomb law. So near the origin, Eq. (6.223) should go over into the Coulomb law, and thus it follows that

$$\chi(t) = de(t),\ \phi = \frac{de(t - R/c)}{R} \tag{6.225}$$

Introducing the charge density $\rho(x, y, z, t)$, we get

$$de = \rho dV$$

The solution of Eq. (6.217) is

$$\phi(x, y, z, t) = \int \frac{1}{R}\rho(x', y', z', t - R/c)dV' + \phi_0 \tag{6.226}$$

where

$$R^2 = (x - x')^2 + (y - y')^2 + (z - z')^2$$
$$dV' = x', y', z', t'$$

and ϕ_0 is the solution of Eq. (6.217) without the right-hand side.
Similarly Eq. (6.216) has the solution

$$\mathbf{A} = \frac{1}{c}\int \frac{\mathbf{J}(x', y', z', t - R/c)}{R}dV' + \mathbf{A}_0 \tag{6.227}$$

where $\mathbf{A}_0$ is the solution of Eq. (6.216) without the right-hand side.

Equations (6.226) and (6.227) take into account the finite velocity c of propagation of the electromagnetic field, since if we desire to determine the potential at a point P at a time t, we have to take the charge density not at P, but at the retarded position P' at the previous time $t - R/c$, where R/c is the time which the signal needs to reach P from its retarded position P'.

6.10.1 The Lienard–Wiechert Potentials

A given charge distribution ρ produces the field given by equations for the retarded potentials, Eqs (6.226) and (6.227) without ϕ_0 and $\mathbf{A}_0$

$$\phi(x, y, z, t) = \int \frac{1}{R}\rho(x', y', z', t - R/c)dV' \tag{6.228}$$

$$\mathbf{A} = \frac{1}{c}\int \frac{\mathbf{J}(x, y, z, t - R/c)}{R}dV' \tag{6.229}$$

where $t' = (t - R/c)$ is the retarded time of the charge in motion. For carrying the transition to a point charge, we will first transform the integrals in Eqs (6.228) and (6.229) into integrals over the charge elements de. We suppose that the charge elements are rigidly connected with each other and have at a given time the same velocity $\mathbf{u}(t')$. If $P(x, y, z, t)$ is the field point where we intend to find the field at time t, consider a spherical shell of thickness dR at a distance R from P, Figure 6.2.

A volume element of this shell will be $dV = d\sigma dR$. The contribution to the integral, Eq. (6.228), is the charge density ρ divided by R, which a spherical light wave, which contracts with velocity c, meets and reaches P at t. The wave will pass the outer surface of the shell at a time t'. During the time $dt = dR/c$, which the light wave takes to pass the shell of thickness dR, an amount of charge that will leak through the inner surface of the shell, is given by

$$-\rho\frac{(\mathbf{u}\cdot\mathbf{R})}{R}dt = -\rho\frac{(\mathbf{u}\cdot\mathbf{R})}{R}\frac{dR}{c} \tag{6.230}$$

where $(\mathbf{uR}) = \mathbf{u}\cdot\mathbf{R}$; the direction of $\mathbf{R}$ is from P to P′ and R is the distance at the retarded time. The charge element de collected by the light wave during the time dt is

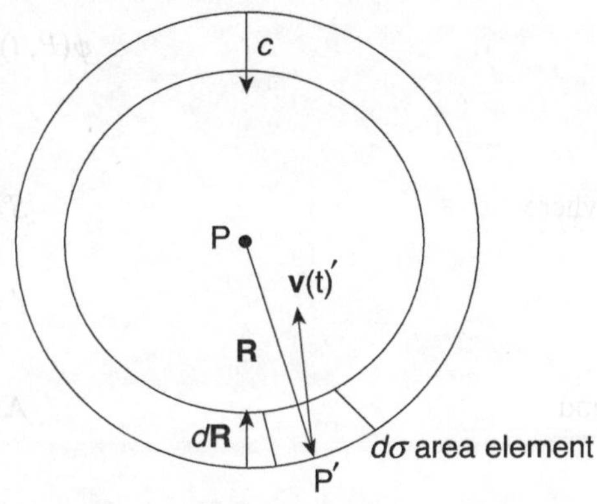

Figure 6.2 *Element dσ of sphere collecting information from charge in motion with velocity c.*

$$de = \rho\left[1 + \frac{(\mathbf{uR})}{Rc}\right]dV \tag{6.231}$$

The integrand of Eq. (6.228) becomes

$$\rho dV = \frac{de}{R + \frac{(\mathbf{uR})}{c}\Big|_{t-R/c}} \tag{6.232}$$

and carrying out the transition to a point charge, we get

$$\phi(P, t) = \frac{e}{R + \frac{(\mathbf{uR})}{c}\Big|_{t-\frac{R}{c}}} \tag{6.233}$$

Analogously

$$\mathbf{A}(P, t) = \frac{1}{c}\frac{e\mathbf{u}}{R + \frac{(\mathbf{uR})}{c}\Big|_{t-\frac{R}{c}}} \tag{6.234}$$

The potentials of the field in the form of Eqs (6.233) and (6.234) are called the Lienard–Wiechert potentials.

6.11 FIELD STRENGTHS OF AN ACCELERATED CHARGE

The field strengths **E** and **B** are calculated from Eqs (6.233) and (6.234) by the relations

$$\mathbf{E} = -\nabla\phi - \frac{1}{c}\frac{\partial \mathbf{A}}{\partial t} \tag{6.235}$$

and

$$\mathbf{B} = \nabla \times \mathbf{A} \tag{6.236}$$

$$\phi(P, t) = \frac{e}{R + \frac{(\mathbf{uR})}{c}\Big|_{t-\frac{R}{c}}} = \frac{e}{S}\Big|_{t'} \tag{6.237}$$

where

$$S = R + \frac{(\mathbf{uR})}{c} \tag{6.238}$$

$$t' = t - \frac{R}{c} \tag{6.239}$$

and

$$\mathbf{A}(P, t) = \frac{1}{c}\frac{e\mathbf{u}}{S}\Big|_{t'} \tag{6.240}$$

Therefore,

$$\mathbf{E} = -\nabla\phi - \frac{1}{c}\frac{\partial \mathbf{A}}{\partial t}$$

$$= -\nabla\left(\frac{e}{S}\right)_{t'} - \frac{1}{c}\frac{\partial}{\partial t}\left(\frac{1}{c}\frac{e\mathbf{u}}{S}\right)_{t'}$$

or

$$\frac{\mathbf{E}}{e} = -\nabla\left(\frac{1}{S}\right)_{t'} - \frac{1}{c^2}\frac{\partial}{\partial t}\left(\frac{\mathbf{u}}{S}\right)_{t'} \tag{6.241}$$

The derivatives have to be taken with respect to the time t and the coordinates at P. But the motion of the particle at the time t' is given by $\mathbf{R}(t')$ and $\mathbf{u}(t') = \partial\mathbf{R}(t')/\partial t'$, so the quantities $\mathbf{R}$ and $\mathbf{u}$ are given as functions of retarded time t'. Therefore we have to express the derivatives with respect to t'.

Thus,

$$\frac{\mathbf{E}}{e} = \frac{1}{S^2}\nabla S - \frac{1}{c^2}\frac{\partial}{\partial S}\left(\frac{\mathbf{u}}{S}\right)\frac{\partial S}{\partial t}$$

$$= \frac{1}{S^2}\nabla S + \frac{\mathbf{u}}{c^2 S^2}\frac{\partial S}{\partial t} - \frac{1}{c^2 S}\frac{\partial \mathbf{u}}{\partial t} \tag{6.242}$$

Similarly,

$$\mathbf{B} = \nabla \times \mathbf{A}$$

$$\frac{c\mathbf{B}}{e} = \nabla \times \frac{\mathbf{u}}{S}$$

$$= \frac{1}{S}(\nabla \times \mathbf{u}) + \nabla\left(\frac{1}{S}\right) \times \mathbf{u}$$

$$= \frac{1}{S}(\nabla \times \mathbf{u}) + \frac{1}{S^2}[\mathbf{u} \times \nabla S]$$

$$= \frac{1}{S}[\nabla\mathbf{u}] + \frac{1}{S^2}[\mathbf{u}\nabla S] \tag{6.243}$$

where $[\nabla\mathbf{u}] = \nabla \times \mathbf{u}$ and $[\mathbf{u}\nabla S] = \mathbf{u} \times \nabla S$.

The retarded time t' is given by the distance R at t'.

$$R(t') = c(t - t') \tag{6.244}$$

so

$$\frac{\partial R}{\partial t} = \frac{\partial R}{\partial t'}\frac{\partial t'}{\partial t} \tag{6.245}$$

Since $R^2 = \mathbf{R}\cdot\mathbf{R}$, differentiating both sides with respect to t', we get

$$2R\frac{\partial R}{\partial t'} = 2\mathbf{R}\cdot\frac{\partial \mathbf{R}}{\partial t'}$$

or

$$\frac{\partial R}{\partial t'} = \frac{(\mathbf{Ru})}{R}$$

Further

$$R(t') = c(t - t')$$

$$\frac{\partial R}{\partial t} = c - c\frac{\partial t'}{\partial t} = c\left(1 - \frac{\partial t'}{\partial t}\right)$$

Substituting these results in Eq. (6.245), we obtain

$$c - c\frac{\partial t'}{\partial t} = \frac{(\mathbf{Ru})}{R}\frac{\partial t}{\partial t'}$$

or

$$\left(c + \frac{(\mathbf{Ru})}{R}\right)\frac{\partial t'}{\partial t} = c$$

Hence,

$$\frac{\partial t'}{\partial t} = \frac{1}{1 + \dfrac{(\mathbf{Ru})}{Rc}} = \frac{R}{S} \tag{6.246}$$

where we made use of the definition of S, $S \equiv R + (\mathbf{Ru})/c$.
From Eq. (6.244), we see that t' is a function of the coordinates of P. So, therefore, we get

$$\nabla t' = \frac{\partial t'}{\partial R(t)} = \frac{\partial t'}{\partial R(t')}\frac{\partial R(t')}{\partial R(t)} \tag{6.247}$$

Now
$$R(t') = ct - ct'$$
$$ct' = ct - R(t')$$

Hence,
$$\frac{\partial t'}{\partial R(t')} = -\frac{1}{c} \tag{6.248}$$

So
$$\nabla t' = -\frac{1}{c}\nabla R(t')$$

or
$$\nabla R(t') = -c\nabla t'$$
$$= \nabla_1 R + \frac{\partial R}{\partial t'}\nabla t'$$

where ∇_1 represents the gradient at constant t'.

Therefore,
$$\nabla R(t') = \frac{\partial R}{\partial t'}\nabla t' - \frac{\mathbf{R}}{R}$$
$$= \frac{(\mathbf{Ru})}{R}\nabla t' - \frac{\mathbf{R}}{R}$$

Hence,
$$\nabla t' = -\frac{1}{c}\nabla R(t')$$
$$= -\frac{1}{c}\left(\frac{(\mathbf{Ru})}{R}\nabla t' - \frac{\mathbf{R}}{R}\right)$$

or
$$\nabla t'\left(1 + \frac{(\mathbf{Ru})}{cR}\right) = \frac{\mathbf{R}}{cR}$$
$$\nabla t' = \frac{\mathbf{R}}{cR}\frac{1}{1 + \dfrac{(\mathbf{Ru})}{cR}}$$
$$= \frac{\mathbf{R}}{cR + (\mathbf{Ru})} = \frac{\mathbf{R}}{cS} \tag{6.249}$$

The derivatives of S are given as
$$\frac{\partial S}{\partial t} = \frac{\partial S}{\partial t'}\frac{\partial t'}{\partial t}$$
$$= \frac{R}{S}\frac{\partial}{\partial t'}\left(R + \frac{(\mathbf{Ru})}{c}\right)$$
$$= \frac{R}{S}\left(\frac{(\mathbf{Ru})}{R} + \frac{u^2}{c} + \frac{(\mathbf{R\dot{u}})}{c}\right) \tag{6.250}$$

and
$$\nabla S=\nabla\left(R+\frac{(\mathbf{Ru})}{c}\right)$$

$$=\nabla_1 S+\frac{\mathbf{R}}{cS}\left(\frac{(\mathbf{Ru})}{R}+\frac{u^2}{c}+\frac{(\mathbf{R\dot{u}})}{c}\right)$$

$$=\frac{\partial}{\partial \mathbf{R}}\left(R+\frac{(\mathbf{Ru})}{c}\right)+\frac{\mathbf{R}}{cS}\left(\frac{(\mathbf{Ru})}{R}+\frac{u^2}{c}+\frac{(\mathbf{R\dot{u}})}{c}\right)$$

or
$$=-\frac{\mathbf{R}}{R}-\frac{\mathbf{u}}{c}+\frac{\mathbf{R}}{cS}\left(\frac{(\mathbf{Ru})}{R}+\frac{u^2}{c}+\frac{(\mathbf{R\dot{u}})}{c}\right) \tag{6.251}$$

because the direction of vector **R** is from P' to P whereas the direction of the vector along which it is being differentiated is in the opposite direction.

Furthermore,
$$\dot{\mathbf{u}}=\frac{\partial \mathbf{u}(t')}{\partial t'}$$

$$\frac{\partial \mathbf{u}}{\partial t}=\frac{\partial \mathbf{u}}{\partial t'}\frac{\partial t'}{\partial t}=\dot{\mathbf{u}}\frac{R}{S} \tag{6.252}$$

$$\nabla t'\times\mathbf{u}=-[\mathbf{u}\nabla t']$$

$$=\frac{[\mathbf{Ru}]}{cS} \tag{6.253}$$

The field strengths are given by Eqs (6.242) and (6.243) and inserting the derivatives of S into these equations, we get

$$\frac{\mathbf{E}}{e}=\frac{1-\beta^2}{S^2}\left[\mathbf{R}+\tfrac{u}{c}\mathbf{R}\right]+\frac{1}{S^3c^2}\left[\mathbf{R}\left[\mathbf{R}+\tfrac{u}{c}\mathbf{R},\,\dot{\mathbf{u}}\right]\right] \tag{6.254}$$

$$\frac{\mathbf{B}}{c}=\frac{[\mathbf{ER}]}{R} \tag{6.255}$$

The magnetic field is perpendicular to the electric field and to **R**, whereas the electric field has two components, one perpendicular to **R** and the other along **R**. The formulae, Eqs (6.254) and (6.255), are valid for any given motion of the particle so far as the particle can be considered a point charge, so that their validity does not hold at distances smaller than the electronic radius.

The electric field equation, Eq. (6.254), consists of two parts. The first part depends upon the velocity and varies as R^{-2} and reduces to $-\mathbf{R}/R^3$ for $u = 0$ and so represents the static (coulomb) part of the field. It represents a non-radiating term of the field in the sense that it does not contribute to the energy flow over an infinitely distant surface. The second part is proportional to **u**, decreases as R^{-1} for long distances and the field is perpendicular to the radius vector **R**. It does represent a radiation field since it contributes to the energy flux over a large sphere. The region where the effect of second part predominates is called the *wave zone*.

Problems

6.1 Verify that the Lagrangian

$$\mathcal{L} = -mc^2\sqrt{1 - u^2/c^2} + \frac{e}{c}\,\mathbf{A}\cdot\mathbf{u} - e\phi$$

yields the correct equation of motion.

6.2 Show that in the non-relativistic approximation the Lagrangian and Hamiltonian reduce to the following forms respectively

$$\mathcal{L} = \frac{mu^2}{2} + \frac{e}{c}\,\mathbf{A}\cdot\mathbf{u} - e\phi$$

$$\mathcal{H} = mc^2 + e\phi + \frac{p^2}{2m} - \frac{e}{mc}\,(\mathbf{P}\cdot\mathbf{A}) + \frac{e_2}{2mc^2}A^2$$

6.3 Find the Hamilton equations of motion of a free particle using cylindrical coordinates r, z, t.

6.4 Show that the gauge transformation of the potentials

$$A'_\mu = A_\mu + \frac{\partial f}{\partial x_\mu}$$

where f is an arbitrary function of the coordinates and time, in the charged particle Lagrangian

$$\mathcal{L} = -mc^2\sqrt{1 - u^2/c^2} + \frac{e}{c}\,\mathbf{A}\cdot\mathbf{u} - e\phi$$

gives another equivalent Lagrangian.

6.5 Show that the equation of motion of a charge in electromagnetic field in the four-dimensional form

$$m\,\frac{du_\mu}{dt} = \frac{e}{c}\sum_{\nu=0}^{3} F_{\mu\nu}u^\nu$$

for $\mu = 1, 2, 3$ reduces to the corresponding form in three-dimensions, i.e.,

$$\mathbf{F} = e\left[\mathbf{E} + \frac{1}{c}(\mathbf{u}\times\mathbf{B})\right]$$

and for $\mu = 0$, it becomes

$$\frac{d}{dt}\left(mc^2/\sqrt{1 - u^2/c^2}\right) = e\mathbf{E}\cdot\mathbf{u}$$

where $\mathbf{u}$ is the velocity of the particle.

6.6 If a particle of charge e moves in a plane perpendicular to a magnetic field $\mathbf{B}$ not varying in time, show that the radius of curvature ρ is given by $eB\rho = cp$, where p is the magnitude of the (relativistic) momentum of the particle. If $\mathbf{B}$ is also constant in space, show that plane motion in a circle is described with angular frequency

$$\omega = \frac{eB}{m_r c} = \frac{eB}{m\beta c}$$

where m is the rest mass of the particle and $\beta = m_r/m$.

6.7 In a certain reference frame, there exist electric and magnetic fields **E** and **B** such that **E** > **B**. Can there exist, another reference frame in which there will be (a) only electric field but no magnetic field, (b) only magnetic induction but no electric field.

6.8 Calculate radius of curvature of 150 MeV electron in a magnetic field of 10,000 gauss, when particle velocity and magnetic field are perpendicular to each other.

6.9 Show that

(a) the equation of continuity expressed in four-dimensional form is

$$\sum_{\mu=0}^{3} \frac{\partial j^{\mu}}{\partial x^{\mu}} = 0$$

(b) Lorentz condition in the four-dimensional form is

$$\sum_{\mu=0}^{3} \frac{\partial A^{\mu}}{\partial x^{\mu}} = 0$$

6.10 Starting from the four-dimensional form of homogeneous Maxwell's equations, viz.,

$$\sum_{\nu} \frac{\partial F^{\mu\nu}}{\partial x^{\nu}} = 0 \qquad (\mu = 0, 1, 2, 3)$$

obtain the wave equation for the field in a vacuum in the four-dimensional form. Further show that the equation reduces to the following equations for the potentials in the absence of charges and currents (i.e., $\rho = 0$ and $\mathbf{j} = 0$), viz.,

$$\nabla^2 \mathbf{A} - \frac{1}{c^2}\frac{\partial^2 \mathbf{A}}{\partial t^2} = 0$$

$$\nabla^2 \phi - \frac{1}{c^2}\frac{\partial^2 \phi}{\partial t^2} = 0$$

6.11 Show that the retarded potentials

$$\phi(P, t) = \int \frac{\rho\left(P', t - R_{pp'}/c\right)}{R_{pp'}} dV'$$

$$\mathbf{A}(P, t) = \frac{1}{c} \int \frac{(\rho \mathbf{u})(P', t - R_{pp'}/c)}{R_{pp'}} dV'$$

satisfy the Lorentz condition

$$\nabla \cdot \mathbf{A} + \frac{1}{c}\frac{\partial \phi}{\partial t} = 0$$

6.12 Prove that electric charge is Lorentz invariant, i.e., it is not altered by motion relative to an observer.

6.13 Show that a system S', in which the field vector E' and $\mathbf{B}'$ turn out to be parallel, is determined by the equation

$$\frac{v/c}{1 + v^2/c^2} = \frac{\mathbf{E} \times \mathbf{B}}{E^2 + B^2}$$

where v is the velocity of the system S' relative to S, **E** and **B** being the intensity vectors in S.

6.14 In projection television sets, electrons are accelerated through a potential difference of 50 kV. Assuming that the electrons start from rest, calculate the speed of the electrons relativistically as well as non-relativistically. Comment on the significance of the difference in speed in the design of set.

(0.413c; 0.442c; no difference in the operation of the set.)

6.15 Show that the momentum of a particle having charge e and moving in a circle of radius R is given by $p = 300\ BR$, where p is in MeV/c, B in T, and R in m. BR is called magnetic rigidity.

6.16 Calculate the radius of curvature of 150 MeV electrons in a magnetic field of 10,000 gauss. The particle velocity and magnetic field are orthogonal to each other. (50 cm)

6.17 A single charged particle describes an arc of 50 cm radius in uniform magnetic field of 10,000 gauss, the velocity and magnetic field are orthogonal to each other. Find magnetic rigidity and momentum when the particle is (*a*) an electron and (*b*) a proton.

[(*a*) 5×10^5 gauss-cm, 150 MeV/c; (b) 5×10^5 gauss-cm, 150 MeV/c]

7

Applications of Special Theory of Relativity

7.1 EXPERIMENTAL VERIFICATION OF THE SPECIAL THEORY[1]

The experimental observation that the velocity of electromagnetic disturbances is the same in frames of reference, which are in uniform translation with respect to each other, was built into the theory by Einstein. In addition, the negative results of Michelson–Morley experiment were explained by Fitzgerald contraction hypothesis, but the theory predicts the contraction of length in the direction of motion. The other predictions of the theory are time dilation, relativistic increase of mass, and the mass energy equivalence. Let us discuss the experimental evidence, which confirms these predictions and thus rivets our faith in this already sound and convincing theory.
There are two possibilities to test special relativity:

1. Testing the postulates
2. Testing the consequences

7.1.1 Testing the Postulates

It is conceptually obvious how to test the postulate of the constancy of velocity of light, in all frames of reference and the recent attempts have been discussed earlier along with the postulates. The second postulate of the relativity principle, viz., there is no way to distinguish between the different frames, can never be tested completely because these are the statements about all physical phenomena. Nonetheless, there are continuous searches for effects violating the relativity principle. Even if there is one physical phenomenon that violates the relativity principle, then special relativity must be wrong.

In order to have a guideline at hand how to make a complete test of special relativity, Robertson (1948) proposed a test theory which singles out three tests, which when given particular results will prove the validity of special relativity. These three tests are as follows:

1. Tests of the isotropy of the velocity of light. In an arbitrary moving frame, the velocity of light is independent of the direction of propagation.
2. Tests of the independence of the velocity of light from the velocity of the source or the laboratory.
3. Tests of the time dilation.

The first tests are called Michelson–Morley (Michelson–Morley, 1887), the second Kennedy–Thorndyke tests (Kennedy and Thorndyke, 1932), and the last Ives–Stilwell tests (Ives and Stilwell, 1938), named after the persons who carried out these tests for the first time.

As the first two tests give a null result and the third a well-known time–dilation effect, the special relativity has been tested to be valid at least within the experimental accuracy.

Until the 1940s, the constancy of the velocity of light was tested with interferometers. Modern tests use optical resonators and lasers. Tests of the Doppler effect use spectroscopic techniques applied to fast atoms or spectral lines.

7.1.2 Testing the Consequences

The consequences of special relativity have been widely tested. Furthermore, the results of special relativity are used in many areas of physics. The high-energy experiments at Large accelerators cannot be interpreted without special relativity.

Furthermore, Ground Positioning System (GPS), telecommunications, satellite ranging, and others are based on the validity of special relativity.

7.1.3 Test of Time Dilatation and Transverse Doppler Effect

The Doppler effect is the observed variation in frequency of a source when it is observed by a detector that is moving relative to the source. This effect is most pronounced when the source is moving directly towards or away from the detector, and in pre-relativity physics its value was zero for transverse motion (perpendicular to the source-detector line). In special relativity, there is a non-zero Doppler effect for transverse motion due to the relative time dilation of the source as seen by the detector. Measurements of Doppler shifts with velocities approaching c can test the validity of special relativity's prediction for such observations, which differs significantly from classical predictions. The experiments support special relativity and are in complete disagreement with non-relativistic predictions. There is a general review article by Gwinner.[2]

The observational evidence of time dilation is overwhelming. Ives and Stilwell[3,4] measured the frequency shifts of radiation emitted in the forward and backward directions by moving ions of H_2 and H_3 molecules where $v/c \sim 6 \times 10^{-3}$. The transformation law for the frequency of light ray from Eq. (5.20) is

$$\nu' = \nu \frac{1 - \cos\theta\, v/c}{\sqrt{1 - v^2/c^2}} \tag{7.1}$$

or

$$\nu' = \nu\left[1 - \cos\theta\, v/c + \frac{1}{2}\left(\frac{v}{c}\right)^2 + \cdots\right]$$

where θ is the angle between the relative velocity of the two coordinate systems and the direction of propagation of light. The first order effect is 'classical' Doppler effect and the second term (independent of θ) is the relativistic effect.

Ives measured the wavelength of H_B emitted by the canal rays in two directions along and against their motion. The undisplaced line of hydrogen atoms at rest was also photographed simultaneously on the same photographic plate, along with these two lines v'_1 and v'_2. The frequencies of displaced lines and from Eq. (5.22) are

$$v'_1 = v\frac{1 - v/c}{\sqrt{1 - v^2/c^2}} \tag{7.2}$$

$$v'_2 = v\frac{1 + v/c}{\sqrt{1 - v^2/c^2}} \tag{7.3}$$

Defining

$$\bar{v} = \frac{v'_1 + v'_2}{2} = \frac{v}{\sqrt{1 - v^2/c^2}} \tag{7.4}$$

and measuring the displacement of $\bar{v}$ with respect to the frequency v, the relativistic transformation law was confirmed. Canal rays act as atomic clocks and it is the slowing down of these clocks that produces the relativistic effect.[4]

When a particle moves with velocity **u**, its proper time τ_0 with reference to an observer is dilated to

$$\tau = \frac{\tau_0}{\sqrt{1 - u^2/c^2}}$$

and its energy $\mathcal{E}$ is given by

$$\mathcal{E} = \frac{\mathcal{E}_0}{\sqrt{1 - u^2/c^2}}$$

Thus,

$$\tau/\mathcal{E} = \tau_0/\mathcal{E}_0 \tag{7.5}$$

This constant ratio between time interval and energy of a free particle has been checked in experiments on high energy particles in both cosmic-rays[5] and that produced artificially[6].

Therefore for a $\mu^+ \rightarrow e^+ + \bar{v}_\mu + v_e$ or $\mu^- \rightarrow e^- + v_e + \bar{v}_\mu$ decay, where v is a neutrino, the lifetime τ_0 for the decay at rest is 2.6×10^{-6} s, but the lifetime is increased corresponding to the energy, if the decay is in flight. Thus the lifetime of a moving meson is dilated in excellent agreement with Eq. (7.5).

More recently, this apparent slowing down of clocks was demonstrated by using Mössbauer effect.[7–8] In an ultracentrifuge to show that an 'observer' moving at right angles to the source (where Doppler effect is excluded) will observe the frequency of monochromatic gamma-rays shifted in accordance with Eq. (2.59).

Another prediction of the theory of relativity is the variation of mass of a particle with its velocity

$$m_r = \frac{m_0}{\sqrt{1 - v^2/c^2}}$$

where m_0 is the mass in its inertial frame. Bucherer (1909) carried out a series of measurements of e/m_r ratio of high velocity electrons of radioactive origin and thus provided the first experimental verification of the relativistic dependence of mass on velocity. Since e, the charge on an electron, being a physical reality, is Lorentz invariant

$$\frac{e}{m_r} = \frac{e}{m_0}\left(1 - \frac{u^2}{c^2}\right)^{1/2} = \left(\frac{e}{m_0}\right)_{\frac{u}{c}\rightarrow 0}\left(1 - \frac{u^2}{c^2}\right)^{1/2} \tag{7.6}$$

where $\left(\frac{e}{m_0}\right)_{\frac{u}{c}\to 0}$ is the value of $\frac{e}{m_r}$ for electrons of velocities smaller than c. Bucherer employed naturally occurring radioactive elements and selected electrons of a certain velocity by the velocity selector arrangements of electric and magnetic fields. These electrons were deflected by magnetic field and e/m_r determined from the deflection. The results obtained were in excellent agreement with predictions given in Eq. (7.6)[9–11].

The rest mass energy of a particle is equal to c^2 times its rest mass, that is, $\mathcal{E}_0 = mc^2$ and the total relativistic energy is equal to c^2 times its relativistic mass, that is,

$$\mathcal{E} = m_r c^2$$

The above two relations between mass and energy, first obtained by Einstein, have been amply verified from the Q-value measurements on nuclear reactions (Section 4.6). The phenomenon of electron–positron pair formation by a gamma-ray of energy greater than 1.02 MeV, the rest-mass energy and the converse of it, and the annihilation of electron–positron pairs into two quanta or three quanta (with or without the presence of matter) lend a quantitative check of these predictions.

It has taken more than a century, but the formula $E = mc^2$ has been finally corroborated, at the scale of subatomic particles, in quantum chromodynamics; thanks to a heroic computational effort by French, German, and Hungarian physicists in estimating the mass of protons and neutrons, using some of the world's mightiest computers. According to the conventional model of particle physics, protons and neutrons comprise smaller particles known as quarks, which in turn are bound by gluons. The odd thing is that the mass of gluons is zero and the mass of quarks is only 5 per cent. Where therefore is the missing 95 per cent? The answer, according to the study published in the US journal *Science*, comes from the movements and interactions of quarks and gluons. In other words, energy and mass are equivalent.

The salient features of relativistic mechanics such as the inertia of energy and momentum flux associated with its flow had acquired the validity of experimental verification in a number of cases. One such case is the Compton effect, in which an incoming x-ray photon suffers an elastic collision with an electron, resulting in the transfer of a part of its energy to the latter and itself getting scattered with an increased wavelength. Theoretical analyses were carried out by Compton[12] and independently by Debye.[13] The expressions for change of wavelength of the photon and the kinetic energy of the recoiling electron, employing the relativistic expressions for the energy and momentum of both the particles and assuming the validity of conservation laws of energy and momentum, are deduced in *Example* 4.9. The agreement[14] between experimental results and theoretical predictions confirmed the belief in the validity of relativistic formulae and conservation laws. Later on, Champion[15] investigated the collisions between fast electrons and electrons practically at rest in a Wilson chamber and showed good agreement between the observations and predictions based on relativistic formulae and the conservation laws. Further support to the correctness of the conservation laws and relativistic dynamics is lent by the agreement between the predicted and experimentally observed values of the threshold energies for the pair productions of proton–antiproton, hyperons, and K-mesons in the meson–nucleons collisions as in *Example* 4.13.

7.1.4 Motivation for Testing Special Relativity

There are three main reasons for performing new and improved tests:

1. Special relativity is one pillar of modern physics. Each basic theory has to be tested at its best because it is so fundamentally important.
2. Test of the validity of Maxwell's equations. Since light is a consequence of Maxwell's equations, each test of special relativity is also a test of the validity of Maxwell's equations.

3. There are predictions from loop gravity (Pullin and Gambini[16] 1999, Alfaro et al.[17] 2000) and string theory (Ellis et al.[18] 1999, Ellis et al.[19] 1999a), giving modified Maxwell and Dirac equations that violate Lorentz invariance. Violations of Lorentz invariance also arise in extensions of the standard model (Colladay and Kostelecki[20] 1997) and in non-commutative field theories (Carrol et al.[21] 2001).

7.1.5 Recent Tests of CPT and Lorentz Invariance[22]

The CPT theorem is a general property of quantum field theories that state (loosely) that any system must behave the same if one applies the CPT transform to it: invert all charges (*C*, charge conjugation), invert all spatial axes (*P*, parity inversion), and invert the direction of time (*T*, time inversion). While one cannot do any of these in the real world, one can perform experiments in which particles are replaced by antiparticles (*C*), one looks at situations in which left and right are interchanged (*P*), and the particles travel along similar path but in opposite directions and have opposite spin polarizations (*T*).

Lorentz invariance is the technical term for the statement that special relativity is valid. Any violation of CPT invariance implies a violation of Lorentz invariance; theories without Lorentz invariance need not have CPT invariance.

7.2 THOMAS PRECESSION: SPIN–ORBIT INTERACTION

Compton first suggested the idea of a spinning electron[23] in 1921 in connection with the unit of magnetism. But the idea found acceptance when Goudsmit and Uhlenbeck[24] proposed in 1925 that an electron has an inherent angular momentum and magnetic moment. This spin angular momentum in the atoms' own magnetic field will give rise to an orientational potential energy, which could be either positive or negative depending upon the direction of spin vector. This additional interaction energy will cause the splitting of energy levels and it thus offered an explanation of anomalous Zeeman effect.

We set to evaluate this orientational potential energy called the spin–orbit energy. An electron with charge $-e$ is revolving around the nucleus of charge $+Ze$, with velocity $\mathbf{u}$ and let $\mathbf{r}$ be the radius of its orbit, Figure 7.1(a). As observed by the observer from the frame of reference of the electron, the nucleus is moving around with velocity $-\mathbf{u}$, Figure 7.1(b).

Then nucleus with charge $+Ze$ moving with velocity $-\mathbf{u}$ gives rise to a current $\mathbf{I} = -Ze\mathbf{u}/c$ which results in a magnetic field at the site of the electron

$$\mathbf{H} = \frac{\mathbf{I} \times \mathbf{r}}{r^3} = -\frac{Ze}{c}\frac{\mathbf{u} \times \mathbf{r}}{r^3} \tag{7.7}$$

Expressing it in terms of an electric field $\mathbf{E}$ acting on the electron

$$\mathbf{E} = +Ze\frac{\mathbf{r}}{r^3} \tag{7.8}$$

Thus according to Coulomb's law

$$\mathbf{H} = -\frac{1}{c}\mathbf{u} \times \mathbf{E}$$

or

$$\mathbf{H} = \frac{\mathbf{E} \times \mathbf{u}}{c} \tag{7.9}$$

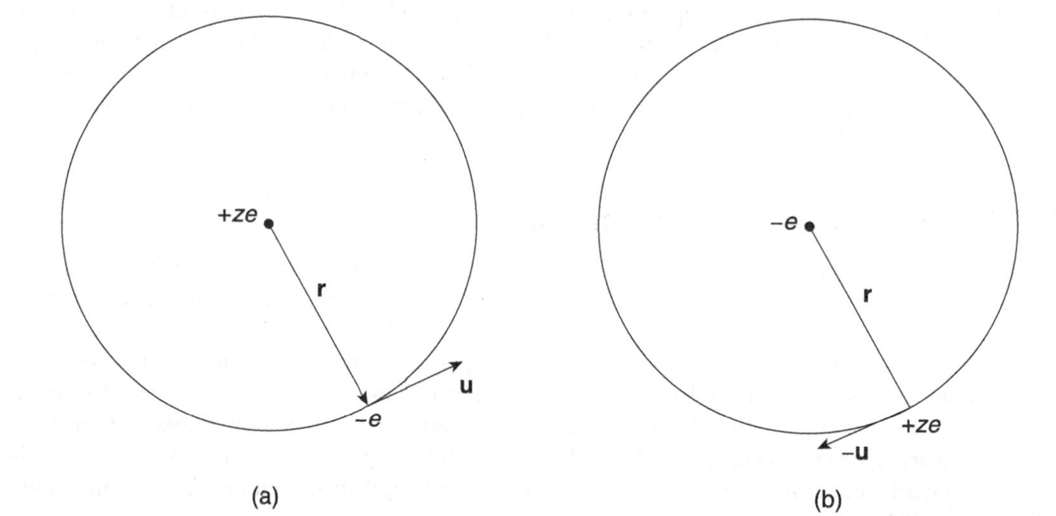

Figure 7.1 *(a) An electron −e revolving around the nucleus +Ze, observed from the inertial. (b) Velocity–u observed from the inertial frame of the electron.*

The magnetic field will cause the spin axis to precess around **H**, with Larmor precession frequency

$$\boldsymbol{\omega}_L = \frac{g_s \mu_B}{\hbar}\mathbf{H} \tag{7.10}$$

where g_s is the Lande's splitting factor for spin; μ_B, the Bohr magneton; and $\hbar = h/2\pi$, the rationalized unit of action.

The spin magnetic moment $\boldsymbol{\mu}_s = -\frac{g_s \mu_B}{\hbar}\mathbf{S}$, placed in this magnetic field will produce an orientational potential energy

$$\Delta\mathcal{E} = -\boldsymbol{\mu}_s \cdot \mathbf{H}$$

$$= \frac{g_s \mu_B}{\hbar}\mathbf{S}\cdot\mathbf{H} \tag{7.11}$$

Equations (7.10) and (7.11) give the Larmor precession frequency and the orientational potential energy in an inertial frame of the electron. But it is not the total spin energy. As the electron moves in the field **E**, it undergoes an acceleration $\boldsymbol{a} = -e\mathbf{E}/m$ and in time dt its velocity changes from **u** to **u** + d**u** with d**u** = $\boldsymbol{a}\,dt$. This continual change of the electron reference frame with respect to the atom reference frame will introduce the additional precession of the spin axis. This kinematical effect of the theory of relativity was first investigated by L. H. Thomas (1927) and is called *Thomas precession.*[25,26]

Let us now calculate this additional precession frequency. X_1Y_1 is a frame of reference, with respect to which the electron is momentarily at rest at the time t_1, similarly in X_2Y_2 it is momentarily at rest at a subsequent time t_2, Figure 7.2. To the observer in frame XY, the axes X_2Y_2 appear slightly rotated through a small angle and those X_3Y_3 rotated still more, although the observer moving with the electron contends that X_3Y_3 is parallel to X_2Y_2 and X_2Y_2 is parallel to X_1Y_1. But as observed by the observer in X_1Y_1, Figure 7.3, the axes XY are moving with velocity $-\mathbf{u}$ in the direction of $-X_2$-axis relative to X_2Y_2 and the electron itself is accelerating towards the nucleus with acceleration $\mathbf{a}$ in the direction of $+Y_2$-axis. The change in velocity is

$$d\mathbf{u} = \mathbf{a}\left(t_1 - t_2\right)$$
$$= \mathbf{a}\, dt \quad (7.12)$$

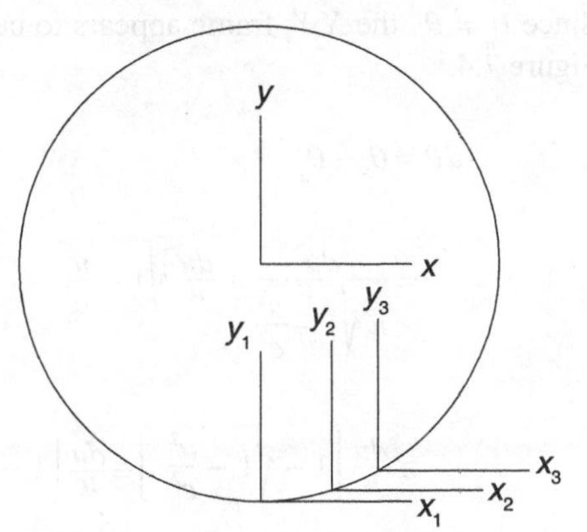

Figure 7.2 *The frame of reference of the electron X_1Y_1 at time t_1 shown at subsequent times t_2 and t_3 as observed from the inertial frame of nucleus XY.*

Equation (7.12) gives the velocity of X_2Y_2 as seen by X_1Y_1. Making use of Eqs (2.70) for the transformation of velocity components, let us evaluate the velocity of X_2Y_2 as seen by XY. Calling $\mathbf{u}_a$ the velocity of X_2Y_2 as seen by XY, we have

$$u_{ax} = \frac{u_x - u}{1 - \frac{uu_x}{c^2}} = \frac{0 + u}{1 - \frac{(-u)(0)}{c^2}} = u \quad [7.13(a)]$$

$$u_{ay} = \frac{u_y\sqrt{1 - \frac{u^2}{c^2}}}{1 - \frac{u_x u}{c^2}} = du\sqrt{1 - \frac{u^2}{c^2}} \quad [7.13(b)]$$

Calling $\mathbf{u}_b$ the velocity of XY as seen by X_2Y_2, we get

$$u_{bx} = \frac{u_x\sqrt{1 - \frac{u^2}{c^2}}}{1 - \frac{u_x u}{c^2}} = -u\sqrt{1 - \frac{u^2}{c^2}} \quad [7.14(a)]$$

$$u_{by} = \frac{u_x}{1 - \frac{u_x u}{c^2}} = -du \quad [7.14(b)]$$

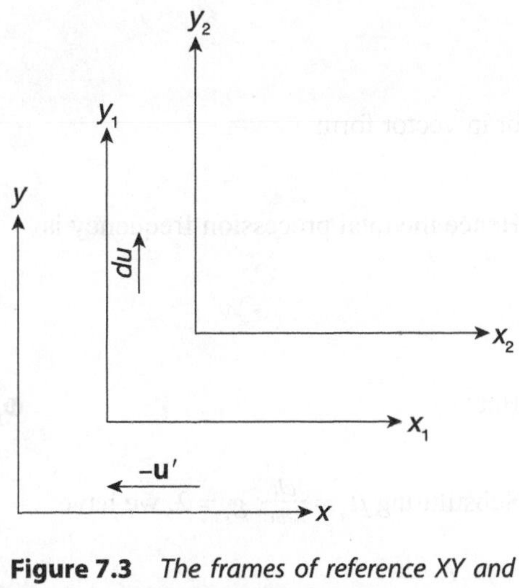

Figure 7.3 *The frames of reference XY and X_2Y_2 as observed from X_1Y_1.*

Now the angle between the vectors u_a and the X-axis of XY frame is

$$\theta_a = \frac{u_{ay}}{u_{ax}} = \frac{du\sqrt{1 - \frac{u^2}{c^2}}}{u} \quad [7.15(a)]$$

and the angle between the vector u_b and the X-axis of X_2Y_2 frame is

$$\theta_b = \frac{u_{by}}{u_{bx}} = \frac{du}{u\sqrt{1 - \frac{u^2}{c^2}}} \quad [7.15(b)]$$

Since $\theta_a \neq \theta_b$, the X_2Y_2 frame appears to be rotated relative to the XY frame and the angle of rotation, Figure 7.4.

$$d\theta = \theta_b - \theta_a$$

$$= \frac{du}{u\sqrt{1-\frac{u^2}{c^2}}} - \frac{du}{u}\sqrt{1-\frac{u^2}{c^2}}$$

$$\cong \frac{du}{u}\left[1-\sqrt{1-\frac{u^2}{c^2}}\right] \cong \frac{du}{u}\left[1-\left[1-\frac{u^2}{2c^2}\right]\right]$$

$$= \frac{duu^2}{2uc^2} = \frac{udu}{2c^2} = \frac{uadt}{2c^2} \tag{7.16}$$

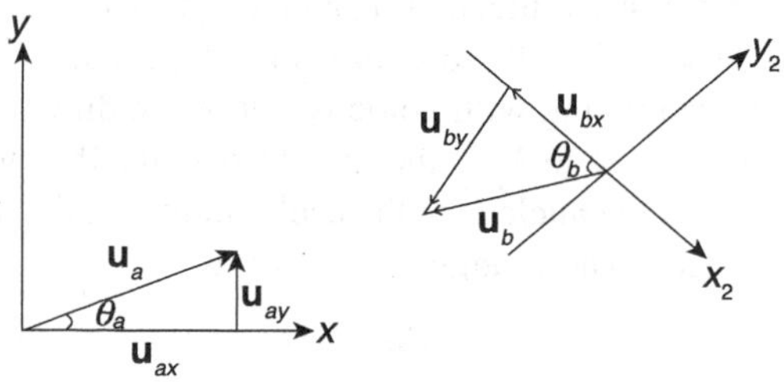

Figure 7.4 *The vectors* $\mathbf{u}_a$ *and* $\mathbf{u}_b$.

Thus the axes X_1Y_1 and X_2Y_2 appear to precess relative to the nucleus with angular frequency ω_T called Thomas frequency.

$$\omega_T = \frac{d\theta}{dt} = \frac{ua}{2c^2}$$

or in vector form

$$\boldsymbol{\omega}_T = -\frac{1}{2c^2}\,\mathbf{u} \times \mathbf{a} \tag{7.17}$$

Hence the total precession frequency is

$$\boldsymbol{\omega} = \boldsymbol{\omega}_L + \boldsymbol{\omega}_T \tag{7.18}$$

But

$$\boldsymbol{\omega}_L = -\frac{g_s\mu_B}{c\hbar}\,\mathbf{u} \times \mathbf{E}$$

Substituting $\mu_B = \frac{e\hbar}{2mc}$; $g_s = 2$, we have

$$\boldsymbol{\omega}_L = -\frac{1}{2c^2}\,\mathbf{u} \times \mathbf{E}$$

and

$$\boldsymbol{\omega}_T = -\frac{1}{2c^2}\,\mathbf{u} \times \mathbf{a}$$

$$= -\frac{1}{2c^2}\,\mathbf{u} \times \left(\frac{-e\mathbf{E}}{m}\right)$$

$$= \frac{1}{2mc^2}\,\mathbf{u} \times \mathbf{E}$$

From Eq. (7.18)

$$\boldsymbol{\omega} = \boldsymbol{\omega}_T + \boldsymbol{\omega}_L$$

$$= \frac{e}{2mc^2}\mathbf{u} \times \mathbf{E} - \frac{e}{mc^2}\mathbf{u} \times \mathbf{E}$$

$$= -\frac{e}{2mc^2}\mathbf{u} \times \mathbf{E} \tag{7.19}$$

From Eq. (7.11), the spin–orbit energy in the frame XY is

$$\Delta\mathcal{E}_{S\text{-}L} = \frac{1}{2}\frac{g_s\mu_B}{\hbar}\mathbf{S}\cdot\mathbf{H} \tag{7.20}$$

From Eq. (7.9), we know

$$\mathbf{H} = -\frac{1}{c}\mathbf{u} \times \mathbf{E}$$

and

$$\mathbf{E} = -\frac{\mathbf{F}}{e}$$

But for any spherically symmetric potential $\phi(r)$

$$\mathbf{F} = -\frac{d\phi(r)}{dr}\frac{\mathbf{r}}{r}$$

Hence,

$$\mathbf{H} = -\frac{1}{c}\mathbf{u} \times \mathbf{E} = \frac{1}{ec}\mathbf{u} \times F$$

$$= -\frac{1}{ecr}\frac{d\phi(r)}{dr}\mathbf{u} \times \mathbf{r}$$

$$= \frac{1}{emcr}\frac{d\phi(r)}{dr}\mathbf{r} \times \mathbf{p}$$

$$= \frac{1}{emcr}\frac{d\phi(r)}{dr}\mathbf{L}$$

From Eq. (7.10)

$$\Delta\mathcal{E}_{S\text{-}L} = \frac{1}{2emc\hbar}\frac{1}{r}\frac{d\phi(r)}{dr}\mathbf{S}\cdot\mathbf{L}$$

Substituting $g_s = 2$; $\mu_B = \frac{e\hbar}{2mc}$, we get

$$\Delta\mathcal{E}_{S.L} = \frac{1}{2e^2c^2}\frac{1}{r}\frac{d\phi(r)}{dr}\,\mathbf{S}\cdot\mathbf{L} \tag{7.21}$$

Although the calculation of the spin–orbit energy as outlined above is essentially non-quantum mechanical, yet the results are in exact agreement with the results of a relativistic correct quantum mechanical calculation.

Example 7.1 Estimate the spin–orbit energy for one-electron atom corresponding to the state of electron in $n = 2$, $l = 1$ state.

Solution.

$$\Delta\mathcal{E}_{S.L} = \frac{1}{2emc\hbar}\frac{1}{r}\frac{d\phi(r)}{dr}\,\mathbf{S}\cdot\mathbf{L}$$

$$\phi(r) = -\frac{e}{r}$$

$$\frac{d\phi(r)}{dr} = \frac{e}{r^2}$$

$$\Delta\mathcal{E}_{S.L} = \frac{e}{2m^2c^2}\frac{1}{r^3}\,\mathbf{S}\cdot\mathbf{L}$$

The average value of $<1/r^3>$ for $n = 2$ state is approximately $1/2\,a_0^3$, where $a_0 = \frac{\hbar_2}{me^2}$. Hence,

$$\Delta\mathcal{E}_{S.L} \approx \frac{e}{2m^2c^2}\frac{m^3e^6}{2^3\hbar^6}\hbar^2$$

$$\approx 10^{-16}\ erg \approx 10^{-4}\ eV$$

But $\mathbf{S}\cdot\mathbf{L}$ can be either positive or negative, depending upon the direction of z-component of spin; the energy level is split by about $2 \times 10^{-4}\ eV$.

The energy level of hydrogen corresponding to $n = 2$, $l = 1$ is $\mathcal{E}_2 = -3.4\ eV$. Thus the splitting is approximately one part in 10^4. This agrees very well with the splitting associated with this level as required to explain the fine structure of the lines.

7.3 RELATIVISTIC CLASSIFICATION OF PARTICLES

For discussing the various types of particles that result from relativistic ideas, we consider the energy momentum four-vector $p^\mu = (\mathcal{E}/c, p_1, p_2, p_3)$. Its scalar product $p^\mu p_\mu$ is given

$$p^\mu p_\mu = \mathcal{E}^2/c^2 - p^2 = m_0^2c^2 \tag{7.22}$$

where m_0 is the rest mass of the particle and is Lorentz invariant.

Taking energy and momentum as the basic variables, we consider the energy-momentum four-vector invariant, in the two-dimensional Minkowski diagram, Figure 7.5.

The three possibilities that result accordingly as

1. $\mathcal{E}^2 - c^2p^2 > 0$
2. $\mathcal{E}^2 - c^2p^2 = 0$
3. $\mathcal{E}^2 - c^2p^2 < 0$

are discussed below.

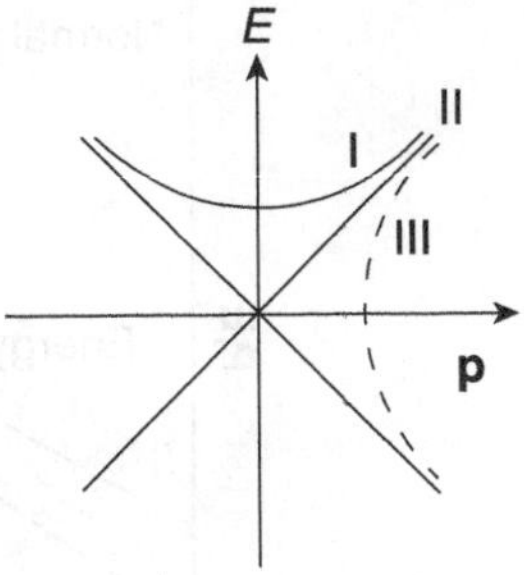

Figure 7.5 *Plot of $(E^2/c^2 - p^2)$ in the $\mathcal{E}$, **p** space.*

(i) $\mathcal{E}^2 - c^2p^2 > 0$.
This case refers to $m_0^2 > 0$, i.e., such particles are represented by hyperbola in the forward light cone region I. The rest mass m_0 is real and positive and these particles always travel with velocity less than c. These are the particles with which we deal in normal life.

(ii) $\mathcal{E}^2 - c^2p^2 = 0$.
It gives $m_0 = 0$ and refers to particles with zero rest mass. These particles travel with the velocity of light and are represented by the degenerate limit of the hyperbola in the upward half of the light cone, marked as region II. Photons and four kinds of neutrinos that have been identified in weak interactions belong to this class.

It may be remarked that the concept of zero rest-mass particle is totally outside the purview of Newtonian mechanics, since the definition of four-momentum

$$p^\mu = m_0 c \frac{dx^\mu}{ds} \tag{7.23}$$

does not hold if $m_0 = 0$. For such particles, the momentum is connected to the energy and not to the velocity and is given by

$$p = \mathcal{E}/c \tag{7.24}$$

Furthermore, for large momentum $p >> m_0c$, $\mathcal{E} = pc$ and such particles behave like zero rest-mass particles.

(iii) $\mathcal{E}^2 - c^2p^2 < 0$
Here m_0^2 is negative and the rest mass for this class of particles is imaginary. These particles travel with a velocity v greater than c. Some physicists had suggested that particles can indeed travel faster than light. According to them, there is no reason why a particle that is already moving at a velocity $v > c$ cannot exist. Tachyons, from the Greek word *tachys* (meaning swift), is the name given to such particles. It is assumed that their rest mass is imaginary, since it is not observable and their energy and momentum are real. It may be remarked that their existence has been postulated in accordance with the theory of relativity.

It is implied that a tachyon, on losing energy, gets speeded up until it is travelling very fast and then it has no energy at all. This characteristic is depicted in Figure 7.6. Some attempts have been made for producing tachyons in the laboratory. At Princeton, to *n*, T. Alväger and M. N. Kriesler expected to produce a pair of equally and oppositely charged tachyons T_+ and T_- by surrounding a source of γ-rays, Cs-134 with some materials. However, no successful detection could be made as a tachyon is supposed to lose its energy in a distance of 10^{-3} cm.

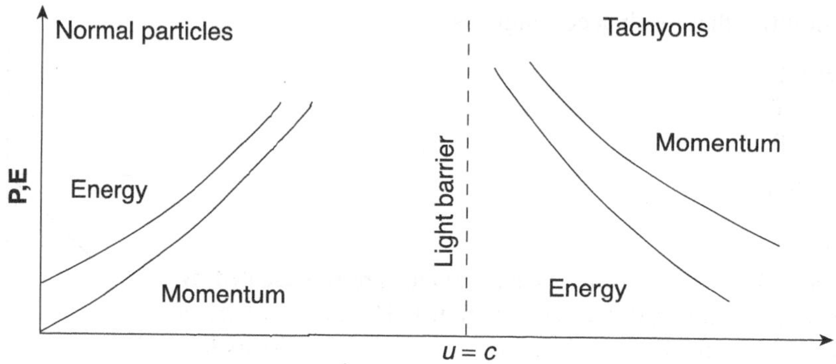

Figure 7.6 *The momentum* **p** *and energy* $\mathcal{E}$ *plotted as a function of velocity. For normal particles, the velocity increases with increasing energy; however, reverse is the case for tachyons.*

7.4 ZERO REST-MASS PARTICLES: PHOTONS AND NEUTRINOS

Light has a dual nature, sometimes behaving as waves and sometimes as corpuscles. The phenomena of interference and diffraction are explained considering the wavy nature of light whereas the photoelectric and Compton effects are explainable only when the corpuscular character is attributed to it. However, the dual nature is a complex momentary phenomenon, that is, we cannot consider light to behave as waves and quanta at the same time. When it is considered as corpuscles, quanta or photons, it is associated with energy $\mathcal{E}$ and momentum **p**; when considered as waves, it is characterized by frequency ν and wavelength λ. We set ourselves the task of finding the relationship between the corpuscular quantities, energy and momentum, and the wave quantities, frequency and wavelength λ. This will be done with the help of special theory of relativity.

The momentum vector **p** and energy $\mathcal{E}$ of a particle are given by

$$\mathbf{p} = m_r\mathbf{u} = \frac{m\mathbf{u}}{\sqrt{1 - v^2/c^2}} \tag{7.25}$$

$$\mathcal{E} = m_r c^2 = \frac{mc^2}{\sqrt{1 - v^2/c^2}} \tag{7.26}$$

This shows that **u** can never be greater than c; even for the limiting value c, the particle will have infinite mass and energy. However, in the case of velocity $u \to c$, if the rest mass tends to zero, the mass and energy will remain finite. Such a particle of zero rest mass is Einstein's light quantum or photon, which moves with the velocity c. Besides the photon, there is another particle of zero rest mass called neutrino (which has an anti-particle called anti-neutrino), which was first postulated theoretically to uphold our faith in the laws of conservation of energy and momentum, in connection with the continuous distribution of electron energy in β-decay process[27]. This remained unobserved for over a quarter of century, but direct experimental evidence for its existence was obtained by Reines and Cowan[28] (1956). This particle differs from the photon in the fact that its spin is $\hbar/2$ whereas that of photon is $\hbar$.

Substituting $u = c$ for a photon, we get

$$p = mu = \frac{\mathcal{E}}{c^2}u = \frac{\mathcal{E}}{c^2}c = \frac{\mathcal{E}}{c} \tag{7.27}$$

If λ is the wavelength and ν, the frequency in the corresponding wave picture, then the wave number k is defined as

$$k = \frac{1}{\bar{\lambda}} = \frac{2\pi\nu}{c} = \frac{\omega}{c} \tag{7.28}$$

Comparing Eqs (7.27) and (7.28), we see that the momentum $\mathbf{p}$ corresponds to the wave number k and the energy $\mathcal{E}$ corresponds to the frequency ν. Since $\mathbf{p}$ is a three-vector, to get the desired relationship between the particle and wave pictures, we therefore assume that $\mathbf{k}$ is also a three-vector with components k_x, k_y, k_z; which determine the direction of propagation of the light waves.

We know that $\mathcal{E}/c, p_x, p_y, p_z$ and $\omega/c, k_x, k_y, k_z$ are four-vectors. So if there exists an invariant relation between the corpuscular quantities $\mathcal{E}, p_x, p_y, p_z$ and the wave theory quantities ω, k_x, k_y, k_z, it must be of the form

$$\frac{\mathcal{E}}{\omega} = \frac{p_x}{k_x} = \frac{p_y}{k_y} = \frac{p_z}{k_z} = \hbar, \text{ say} \tag{7.29}$$

Since the relation (7.29) has to be independent of the choice of coordinate axes, h must be an absolute constant. It is called Planck's constant since Max Planck first introduced it in connection with black body radiation.

Thus from Eq. (7.29) we have

$$\mathcal{E} = \hbar\omega \qquad [7.30(a)]$$

$$p_x = \hbar k_x \qquad [7.30(b)]$$

$$p_y = \hbar k_y \qquad [7.30(c)]$$

$$p_z = \hbar k_z \qquad [7.30(d)]$$

or

$$p = \hbar k = \frac{h}{\lambda} = \frac{h\nu}{c} \tag{7.31}$$

Thus, Eq. (7.31) is consistent with Eq. (7.27). The photon has energy $h\nu$ and momentum $h\nu/c$ and the same ray of light when considered as a wave has frequency ν and wavelength λ, with energy $h\nu$ and momentum $h\nu/c$.

7.4.1 Parity Non-conservation in Weak Interactions

A system of n particles is characterized by a wave function $\psi(\mathbf{r}_1, \mathbf{r}_2, \ldots, \mathbf{r}_n)$ and parity operation $\hat{P}$ is defined as reflecting the coordinate system which describes it in the origin, that is, replacing (x, y, z) by $(-x, -y, -z)$. Thus this operation is equivalent to changing from a right-handed system to a left-handed system of coordinate axes. Writing it in symbols, we have

$$\mathrm{P}\psi(\mathbf{r}_1, \mathbf{r}_2, \ldots, \mathbf{r}_n) = \psi(-\mathbf{r}_1, -\mathbf{r}_2, \ldots, -\mathbf{r}_n) \tag{7.32}$$

Repeating the operation, we get

$$P^2\psi(\mathbf{r}_1, \mathbf{r}_2, \ldots, \mathbf{r}_n) = \psi(\mathbf{r}_1, \mathbf{r}_2, \ldots, \mathbf{r}_n) \tag{7.33}$$

Thus, $\hat{P}^2$ has the eigenvalues ±1, so the eigenvalues of $\hat{P}$ are +1 or −1. We construe that eigen functions of the inversion operator are either unchanged or change in sign when acted upon by $\hat{P}$. The wave function is said to be even in the former case and odd in the latter.

Let us find the conditions under which parity is a constant of motion. When the inversion transformation is applied to scalar quantities, either they do not change at all (true scalars) or they change sign (pseudo-scalars). If a physical quantity is a true scalar, its operator $\hat{f}$ commutes with $\hat{P}$ such that

$$\hat{p}\hat{f} - \hat{f}\hat{p} = 0 \tag{7.34}$$

Equation (7.34) means that if the matrix of $\hat{P}$ is diagonal, then the matrix of $\hat{f}$ is diagonal with respect to the suffix which shows the parity of the state, that is, only those matrix elements which correspond to transitions from even–even or odd–odd are non-zero. The necessary condition for the wave function to have definite parity is that the energy operator $\hat{H}$ of the system is unchanged under coordinate reflection. The invariance of $\hat{H}$ with respect to inversion

$$\hat{H}\hat{p} - \hat{p}\hat{H} = 0 \tag{7.35}$$

The above equation expresses the law of conservation of parity, that is, if the parity of the state of a closed system is even or odd, then it remains even or odd. This will be case only when $\hat{H}$ contains only scalar but not pseudo-scalar terms. Parity conservation holds to an accuracy of at least $1{:}10^{-8}$ in nuclear and electromagnetic interactions and so the energy operator in these cases contains only true scalar terms.

Lee and Yang[29] in 1956 proposed that the parity may not be conserved in weak interactions, but the first experiment which established parity non-conservation in beta-decay was carried out by Wu[30] in 1957, by demonstrating that the angular distribution of electrons (e) from polarized Co^{60} nuclei was anisotropic. This observation that the beta particles which are emitted from nuclei are polarized[31] causes the conventional concept to be overthrown. A qualitative explanation of the non-conservation of parity may be offered as follows.

In a normal beta-decay experiment, we measure the momentum of the electron $\mathbf{p}_e$ and by means of the recoil of the nucleus, the neutrino momentum $\mathbf{P}_\nu$. The only combination of these vectors is $\mathbf{p}_e \cdot \mathbf{p}_\nu$ which is a scalar. But if the beta-emitting nuclei are polarized to begin with, that is, their spins vectors **I** are turned in a direction, then we have this vector enter our measurements. Now vector **I** is due to the orbital angular momentum and the spin which itself is the angular momentum (in the inertial frame of the nucleus), so $\mathbf{r} \times \mathbf{p}$ is an axial vector. If a term of the form $\mathbf{I} \cdot \mathbf{P}_e$ occurs in the beta interaction, the latter will not be invariant under parity operation.

Experimentally the angle θ between **I** and $\mathbf{P}_e$ of the electron decaying from polarized nuclei of Co^{60} was measured and a very definite asymmetry of 30% was observed. Now the inversion transformation changes $\mathbf{P}_e \to -\mathbf{P}_e$, $\mathbf{I} \to \mathbf{I}$, and $\theta \to \theta - \pi$. Hence if parity is conserved, the electron emission should be symmetric about a plane perpendicular to **I**. But as it was not the case, so parity is not conserved in weak interaction of beta-decay. Later on, it was also shown that the weak interaction of $\mu^{\pm} \to e^{\pm} + \nu$ decay and the decay of K-meson do not conserve the parity (θ, τ puzzle).

Example 7.2 Show that the possible explanation of the breakdown of parity conservation in beta-decay does not lie in the zero rest mass of the neutrino.

Solution. A spinning particle in space has two vectors **u** and $\boldsymbol{\sigma}$ attached to it. The possible combination of $\boldsymbol{\sigma}\cdot\mathbf{u}$, which is a pseudo-scalar, would determine the definite parity of the particle. But this is not true for every particle since there is arbitrariness about vector **u**, when viewed by an observer moving relative to it. It could appear to move backward or forward depending upon whether the observer is moving at greater or lesser speed in the same direction. Thus the product $\boldsymbol{\sigma}\cdot\mathbf{u}$ does not have a definite sign at any moment. Since neutrino moves with velocity of light *c*, which is invariant under Lorentz transformation, the product $\boldsymbol{\sigma}\cdot\mathbf{u}$ has a definite sign. It can be shown that $\boldsymbol{\sigma}$ and **u** are always parallel, so neutrino moves in a screw motion of definite right-handed or left-handed parity. But inversion transformation will change the sense of this screw motion, so an isolated particle of zero mass will violate parity conservation. But the overall parity still could be preserved if in any beta-decay process as many left-handed neutrinos are emitted as the right- handed ones. But the experiments do not uphold this equivalence in the number of emitted neutrinos and anti-neutrinos. The force active in β-decay and μ-decay has come to be called the weak interaction.

But the fact that the zero rest mass of neutrino is not the possible cause of parity violation comes into focus when we consider the $\theta - \tau$ puzzle, which was the origin of the conjecture of Lee and Yang that parity is not a valid concept in decay processes. The puzzle is that if the standard laws of conservation and symmetry are accepted, then the τ-meson ($\tau^{\pm} \rightarrow \pi^{\pm} + \pi^{-} + \pi^{\pm}$) cannot have the same spin and parity as the θ-meson ($\theta^{\pm} \rightarrow \pi^{\pm} + \pi^{0}$), as was first shown by Dalitz[32]. From the other experimental data about these particles, we know that their masses are extremely close to each other and the lifetimes agree within the experimental error, of $\sim\pm 5\%$. Furthermore, the absence of any detectable difference between the θ- and τ-mesons in their behaviour with regard to strong interactions was shown by Biswas et al[33]. A number of attempts were made to solve this puzzle, but it was in the early part of 1956 that it seemed that the true solution of the puzzle might lie in the non-conservation of parity[34] in weak interactions. As no neutrinos are involved in these decays, so the zero rest mass of neutrino has no possible bearing on parity non-conservation in case of τ–θ puzzle.

7.4.2 Moving Faster Than Light

There are some arguments that show why an object or even information moving faster than light contradicts the laws of special relativity. Let us describe one such case, Figure 7.7.

We distinguish between three types of lines in time–space:

1. A line with a slope of 1 or –1 (45°) is called a light-like and describes the motion of a light pulse.
2. A line with a slope greater than 1 or smaller than –1, is called a time-like line and can be the *ct*-axis of another frame.
3. A line with a slope greater than –1 and less than 1 is called space-like line and can be the *x*-axis of another frame.

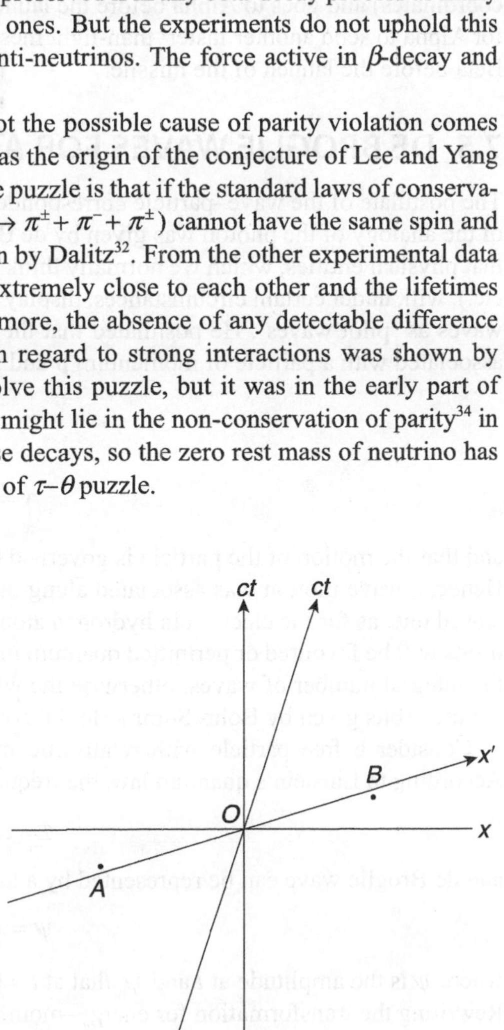

Figure 7.7 *Moving faster than light.*

A line in space-time remains the same in any reference frame. Moving faster than light means that moving along a space-like line in time–space. But two events on a space-like line can happen in opposite order in two frames. In Figure 7.7 we can see that for x, ct coordinate system A happens before B, but for x', ct' coordinate system B happens before A. If we could send information about B along the space-like line connecting it to A in the x', ct' frame, this information would go back in time in the x, ct frame, and get to A before B has happened. This leads to the paradox of action affecting its own cause, like the known example of the time traveller who kills his own grandfather.

Suppose Alpha and Beta are two enemy planes. Event B represents the launch of a missile from Beta that aims to destroy Alpha. A space ship passing by notices the launch and transmits a warning message to Alpha. The message goes faster than light relative to the space ship's frame (x', ct' coordinates) and goes to Alpha before the launch occurred on its own (x, ct). This leaves enough time for Alpha to send another faster–than-light message to a war space ship near Beta that in turn destroys Beta before the launch of the missile.

7.5 DE BROGLIE WAVES FOR A FREE PARTICLE

The postulate of the wave–particle correspondence for particles of non-zero rest mass in continuation of the analogy of the photon was given by de Broglie[35] (1923) in his doctoral dissertation. He thought that physical entities, which we normally think of as particles (electrons, alpha particles, billiard balls, etc.), will, under certain circumstances, display properties characteristic of waves. He designated these waves as 'pilot waves'. He postulated that the wavelength λ and the frequency ν of the pilot waves associated with a particle of momentum $\mathbf{p}$ and total energy $\mathcal{E}$ are given by the equations

$$\lambda = \frac{h}{p} \tag{7.36}$$

$$\nu = \frac{\mathcal{E}}{h} \tag{7.37}$$

and that the motion of the particle is governed by the wave propagation properties of the pilot waves. Hence, a wave motion was associated along the path of a moving particle. If the path of a particle is closed one, as for the electron in hydrogen atom, the waves will interfere with themselves. Only those orbits will be favoured or permitted quantum mechanically, for which the length of the path is equal to the integral number of waves, otherwise the waves will interfere destructively. Such permitted orbits are the orbits given by Bohr–Sommerfeld theory.

Consider a free particle with relativistic mass m_r, rest mass m, and total relativistic energy E. According to Einstein's quantum law, the frequency associated with the particle in its inertial frame is

$$\mathcal{E} = h\nu = mc^2 \tag{7.38}$$

and de Broglie wave can be represented by a harmonic function

$$\psi = \psi_0 e^{2\pi i\nu t} \tag{7.39}$$

where ψ is the amplitude at t and ψ_0 that at $t = 0$.

Rewriting the transformation for energy–momentum four-vector, we have

$$\mathcal{E}' = \frac{\mathcal{E} - up_x}{\sqrt{1 - v^2/c^2}} \tag{7.40(a)}$$

$$p'_x = \frac{p_x - u\mathcal{E}/c^2}{\sqrt{1 - v^2/c^2}} \qquad [7.40(b)]$$

$$p'_y = p_y \qquad [7.40(c)]$$

$$p'_z = p_z \qquad [7.40(d)]$$

Let us transform the energy $\mathcal{E}$ and momentum **p** to the rest inertial frame. Thus,

$$\mathcal{E}' = \frac{mc^2}{\sqrt{1 - v^2/c^2}} \qquad [7.41(a)]$$

$$p'_x = \frac{mu}{\sqrt{1 - v^2/c^2}} \qquad [7.41(b)]$$

$$p'_y = 0 \qquad [7.41(c)]$$

$$p'_z = 0 \qquad [7.41(d)]$$

Further, we transform the wave amplitude ψ, assuming that it is a scalar. Now ψ in the frame S' is given by

$$\psi' = \psi_0 \exp\left[2\pi i\nu \frac{t' + u/c^2}{\sqrt{1 - u^2/c^2}}\right] \qquad (7.42)$$

and its frequency and wavelength are

$$\nu' = \frac{\nu}{\sqrt{1 - u^2/c^2}} = \frac{mc^2/h}{\sqrt{1 - u^2/c^2}} = \frac{\mathcal{E}'}{h} \qquad (7.43)$$

$$\lambda' = \frac{c^2}{u\nu}\sqrt{1 - \frac{u^2}{c^2}} = \frac{h}{mu}\sqrt{1 - \frac{u^2}{c^2}} = \frac{h}{p'} \qquad (7.44)$$

The velocity of the de Broglie wave is given by

$$v_\omega = \nu'\lambda' = \frac{\mathcal{E}}{h}\frac{h}{p'} = \frac{\mathcal{E}}{p'} = \frac{c^2}{u}. \qquad (7.45)$$

The velocity $v_{\omega'}$, called the phase velocity, is greater than c because its product with u equals c^2. However, this is not in contradiction to the basic postulate of special theory, since the phase velocity does not govern the transmission of energy.

When the de Broglie wave is not quite harmonic but composed of two harmonic waves of nearly identical frequencies, the amplitude of the resultant wave is not constant and the variation of it is propagated through space with a velocity v_g called the group velocity and is given by Eq. [5.5(a)]

$$v_g = \frac{d\nu}{d\left(\frac{1}{\lambda}\right)}$$

For a de Broglie wave $1/\lambda = p/h$ and $\nu = \mathcal{E}/h$. Hence, we have

$$v_g = \frac{d\mathcal{E}}{dp} \tag{7.46}$$

But

$$\mathcal{E}^2 - p^2c^2 = m^2c^4$$

or

$$\mathcal{E} = \sqrt{m^2c^4 + p^2c^2}$$

Hence,

$$\frac{d\mathcal{E}}{dp} = \frac{pc^2}{\mathcal{E}} = u \tag{7.47}$$

which is the speed of the particle; thus the group velocity of a de Broglie wave equals the velocity of the particle.

Example 7.3 Consider the motion of a dust particle of radius $r = 10$ cm, $\rho = 10$ g/cm^3, and moving with velocity $u = 10$ cm/s. Show that de Broglie wavelength of this particle is too small compared with its dimensions and thus does not either prove or disprove the validity of de Broglie postulate.

Solution. We have $u = 10$ cm/s and since $u << c$, we can estimate **p** from the classical result

$$p = mu = \frac{4}{3}\pi r^3 \rho u \cong 4 \times 10^{-10} \text{ g-cm/s}$$

The wavelength

$$\lambda = \frac{h}{p} = \frac{6.624 \times 10^{-27}}{4 \times 10^{-10}} \cong 1.7 \times 10^{-17} \text{ cm}$$

which is obviously too small to be observed. Thus the statement that the motion of macroscopic particles does not either prove or disprove the de Broglie postulate, holds.

Example 7.4 Bohr introduced the concept of non-radiating orbits or stationary orbits, where the electron revolving in an orbit around the nucleus does not radiate. Show how it can be reconciled with the standing wave aspect of the electron motion and will lead naturally to the Bohr quantization rule of angular momentum.

Solution. If r is the radius of the orbit, the wave pattern will be a standing wave provided

$$2\pi r = n\lambda = \frac{nh}{p}$$

or

$$rp = n\frac{h}{2\pi} = n\hbar$$

Thus the wave mechanical picture leads to Bohr's postulate that the angular momentum is equal to an integral multiple of $\hbar$. The concept of the electron as a standing wave also satisfies the condition of non-radiating orbits.

7.6 THE UNCERTAINTY PRINCIPLE

The wave particle correspondence is dictated by the special theory of relativity and the standing wave description of an electron in a Bohr orbit is a remarkable idea. The wave packet corresponding to the electron moves with particle velocity, Eq. (7.47). But from the bandwidth theorem of wave motion, we know that shorter the duration of wave motion, the greater is the spread in frequency. This implies that greater the precision with which we determine the position of the electron, the greater is the uncertainty in the determination of momentum and vice versa. This constitutes a fundamental concept of quantum mechanics and an article of faith with the physicists in the description of microscopic phenomena. This uncertainty in the simultaneous measurement of any two canonically conjugate quantities is an unavoidable consequence of the very process of measurement and cannot be got rid of, no matter how accurate the measurement and whichever the method employed.

The wave of a fixed frequency associated with a particle is analysed into a Fourier integral (assuming a continuous frequency spectrum) according to

$$\psi(z) = \int_{-\infty}^{\infty} A(x)e^{2\pi i\nu(t-x/k)}d\nu \tag{7.48}$$

$$= \int_{-\infty}^{\infty} A(\nu)e^{2\pi i\nu z}d\nu \quad z \equiv t - x/c$$

where the expansion coefficient $A(\nu)$ is

$$A(\nu) = \int_{-\infty}^{\infty} \psi(z)\, e^{-2\pi i\nu z}dz \tag{7.49}$$

From Eqs (7.48) and (7.49), we get

$$A(\nu') = \int_{-\infty}^{\infty} \psi(z)\, e^{-2\pi i\nu z}dz$$

$$= \int_{-\infty}^{\infty}\int_{-\infty}^{\infty} A(\nu)\, e^{2\pi i(\nu-\nu')z}d\nu dz \tag{7.50}$$

which shows the existence of the relation

$$\int_{-\infty}^{\infty} e^{-2\pi i(\nu-\nu')z}dz = \delta(\nu-\nu') \tag{7.51}$$

where $\delta(\nu-\nu')$ is the Dirac delta function, defined in such a way that

$$\int_{-\infty}^{\infty} A(\nu)\delta(\nu-\nu')d\nu = A(\nu') \tag{7.52}$$

The expectation or average values of z and ν, are given by

$$<z> = \frac{\int_{-\infty}^{\infty} |\psi|^2 z dz}{\int_{-\infty}^{\infty} |\psi|^2 dz} \tag{7.53}$$

$$<\nu> = \frac{\int_{-\infty}^{\infty} |A|^2 \nu d\nu}{\int_{-\infty}^{\infty} |A|^2 d\nu} \tag{7.54}$$

When ψ is normalized in such a way that $\int_{-\infty}^{\infty} |\psi|^2 dz = 1$, the above definitions become

$$<z> = \int_{-\infty}^{\infty} \psi^2 z dz \tag{7.55}$$

$$<\nu> = \int_{-\infty}^{\infty} A^2 \nu d\nu \tag{7.56}$$

From Eqs (7.49) and (7.51), we have

$$\int_{-\infty}^{\infty} AA^* d\nu = \int_{-\infty}^{\infty}\int_{-\infty}^{\infty}\int_{-\infty}^{\infty} \psi(z)\ \psi^*(z')e^{-2\pi i\nu(z-z')}\, dzdz'd\nu$$

$$= \int_{-\infty}^{\infty}\int_{-\infty}^{\infty} \psi(z)\ \psi^*(z')\ \delta(z-z')\ dzdz'$$

$$= \int_{-\infty}^{\infty} \psi(z)\ \psi^*(z')dz = 1 \tag{7.57}$$

where ψ^* denotes the complex conjugate of ψ
Denoting

$$\Delta z = <(z - <z>)^2>^{1/2} \tag{7.58}$$

$$\Delta \nu = <(\nu - <\nu>)^2>^{1/2} \tag{7.59}$$

the uncertainties in z and ν in analogy with the standard deviation of statistics, we get

$$<(\Delta z)^2> = \int_{-\infty}^{\infty} \psi\psi^*(\Delta z)^2 dz \tag{7.60}$$

$$<(\Delta \nu)^2> = \int_{-\infty}^{\infty} AA^*(\Delta \nu)^2 d\nu \tag{7.61}$$

We shall first present the proof for a wave function for which $<z> = <\nu> = 0$, although the result is generalizable to the case for finite $<z>$ and $<\nu>$. From Eq. (7.48), we have

$$\frac{d^2\psi}{dz^2} = -4\pi^2 \int_{-\infty}^{\infty} \nu^2 A e^{2\pi i\nu z} d\nu$$

which gives, making use of Eq. (7.48), the second time

$$\int_{-\infty}^{\infty} \psi^* \frac{d^2\psi}{dz^2} dz = -4\pi^2 \int_{-\infty}^{\infty}\int_{-\infty}^{\infty}\int_{-\infty}^{\infty} v^2 A(v) e^{2\pi i(v-v')z} A^*(v') dv' dz$$

$$= -4\pi^2 <v^2> \tag{7.62}$$

Integrating by parts and noting that the integrated part vanishes at $-\infty$ and $+\infty$, we find

$$4\pi^2 <v^2> = \int_{-\infty}^{\infty} \frac{d\psi^*}{dz}\frac{d\psi}{dz} dz \tag{7.63}$$

Making use of Schwarz inequality which is the generalization to integral form of the inequality

$$\sum |a_n b_n|^2 \le \sum |a_n|^2 \sum |b_n|^2 \tag{7.64}$$

Also considering the case when the sum splits into two terms, each of which is itself an integral, we have

$$\left| \int_{-\infty}^{\infty} f_1 g_1 dz + \int_{-\infty}^{\infty} f_2 g_2 dz \right|^2 \le \left(\int_{-\infty}^{\infty} |f_1|^2 + |f_2|^2 dz \right) \int_{-\infty}^{\infty} |g_1|^2 + |g_2|^2 dz \tag{7.65}$$

Substituting

$$f_1 = z\psi \tag{7.66}$$

$$f_2 = f^*_1 = z\psi^* \tag{7.67}$$

$$g_1 = \frac{d\psi^*}{dz} \tag{7.68}$$

$$g_2 = g^*_1 \frac{d\psi}{dz} \tag{7.69}$$

We have

$$4\left(\int_{-\infty}^{\infty} z^2 \psi\psi^* dz \right)\left(\int_{-\infty}^{\infty} \frac{d\psi}{dz}\frac{d\psi^*}{dz} dz \right) \ge \left| \int_{-\infty}^{\infty} z\left(\psi \frac{d\psi^*}{dz} + \psi^* \frac{d\psi}{dz} \right) dz \right|^2$$

$$\left| \int_{-\infty}^{\infty} z \frac{d}{dz}(\psi\psi^*) dz \right|^2 = \left| -\int_{-\infty}^{\infty} \psi\psi^* dz \right|^2 = 1$$

Hence

$$16\pi^2 <z^2><v^2> \ge 1 \tag{7.70}$$

But when

$$<z> = <v> = 0 \tag{7.71}$$

$$<\Delta z^2> = <z^2> \tag{7.72}$$

and $$<\Delta v^2> = <v^2> \tag{7.73}$$

Hence, $$16\pi^2(\Delta z^2)(\Delta v^2) \geq 1 \tag{7.74}$$

Writing Δq, Δp for the root mean square values of error in determination of position $\Delta q = v\Delta z$ and momentum $\Delta p = h\Delta v/v$, then

$$\Delta q\, \Delta p \geq \frac{h}{4\pi}$$

or $$\Delta q\, \Delta p \geq \frac{\hbar}{2} \tag{7.75}$$

Substituting $$\Delta z = \Delta t$$

and $$h(\Delta v) = \Delta E$$

we get $$\Delta t\, \Delta E \geq \frac{\hbar}{2} \tag{7.76}$$

This establishes the proof of the uncertainty relation.[36]

Example 7.5 A free electron possesses a magnetic moment $e\hbar/2mc$ and show that it can never be observed directly, say by the use of a magnetometer.[37]

Solution. Suppose that the free electron (whose spin is under investigation) is at a distance r from the magnetometer. Then the magnetic field due to the spin moment is of the order $\frac{e\hbar}{mcr^3}$ and the magnetic field due to the motion of the electron is of the order $\frac{ev}{cr^2}$. For the spin magnetic field to be observable, it must be greater than the uncertainty in the field due to the electronic motion so that we must have

$$\frac{\hbar}{mr} > \Delta v$$

or

$$\frac{\hbar}{m\Delta v} > r$$

where Δv is the uncertainty in our knowledge of **v**. But from uncertainty principle, we know that the uncertainties of our knowledge of r and v are related by

$$\Delta r \Delta v > \frac{\hbar}{m}$$

Making use of this result in the above equation, we get $\Delta r > r$, that is, the uncertainty in the position of the electron under investigation will exceed the distance of electron from magnetometer and so such an experiment will be an impossibility.

Example 7.6 Use Heisenberg uncertainty relation to estimate the ground state energy of harmonic oscillator.

Solution. The energy of harmonic oscillator in a stationary state is

$$\mathcal{E} = \int \psi^*(x) \left\{ \frac{\hat{p}^2}{2\mu} + \frac{k\hat{x}^2}{2} \right\} \psi(x)dx$$

$$= \frac{\overline{\hat{p}^2}}{2\mu} + \frac{k\overline{\hat{x}^2}}{2}$$

Since

$$\overline{\hat{p}^2} = \overline{(\hat{p} - \overline{\hat{p}})^2} + (\overline{\hat{p}})^2$$

$$= \overline{(\Delta\hat{p})^2} + (\overline{\hat{p}})^2$$

$$\overline{\hat{x}^2} = \overline{(\Delta\hat{x})^2} + (\overline{\hat{x}})^2$$

and

$$\overline{\hat{p}^2} = 0 \; ; \overline{\hat{x}^2} = 0$$

Therefore

$$\mathcal{E} = \frac{\overline{(\Delta\hat{p})^2}}{2\mu} + \frac{k\overline{(\Delta\hat{x})^2}}{2}$$

From the Heisenberg relation

$$\overline{(\Delta\hat{p})^2}\;\overline{(\Delta\hat{x})^2} \geq \frac{\hbar^2}{4}$$

it follows that

$$\mathcal{E} \geq \frac{\hbar^2}{8\mu\overline{(\Delta\hat{x})^2}} + \frac{k\overline{(\Delta\hat{x})^2}}{2}$$

The minimization of the above expression leads to

$$\overline{(\Delta\hat{x})^2} = \frac{\hbar}{2}\sqrt{\frac{1}{\mu k}}$$

and we get

$$\mathcal{E}_{min} \sim \frac{\hbar}{2}\sqrt{\frac{k}{\mu}} = \frac{\hbar\omega}{2}$$

where $\omega = \sqrt{\frac{k}{\mu}}$ is the oscillator frequency.

Example 7.7 Ir^{191m} is a gamma-ray emitter and the natural line width of the excited state was found to be $4 \times 10^{-6}\, eV$ by Rudolf Mössbauer[38]. Compute the half-life of the state.

Solution. From the uncertainty relation, we have

$$\Delta t \Delta \mathcal{E} \geq \frac{h}{4\pi}$$

Substituting

$$\Delta \mathcal{E} = \Gamma \text{ (the line width of the state)}$$

$$\Delta t = \tau \text{ (the mean life)}$$

we get

$$\tau \cong \frac{h}{4\pi\Gamma}$$

But

$$\tau_{1/2} = \log 2\tau = 0.692\tau$$

where $\tau_{1/2}$ is the half-life of the state.
Hence

$$\tau_{1/2} = \frac{0.692h}{4\pi\Gamma} = \frac{0.692 \times 6.625 \times 10^{-27}}{4 \times 3.14 \times 4 \times 10^{-6} \times 1.6 \times 10^{-27}}$$

$$= 5 \times 10^{-10}\, s$$

so

$$\tau_{1/2} \sim 10^{-10}\, s.$$

REFERENCES

1. R. Lennuier, "Quelques verifications experimentales recentes de la relativite resreinte", *Rev. Sci., Paris*, (1947) *85*, 740–748. A good survey with main results.

2. G. Gwinner, "Experimental tests of Time dilation in special relativity", *Mod. Phys. Lett.* (2005) *120*, 791.

3. H. F. Ives and G. R. Stilwell, *J. Opt. Soc. Am.* (1938) *28*, 215–226. Also see criticism by R. C. Jones, *J. Opt. Soc. Am.* (1939) *29*, 337–339 and reply by H. F. Ives and G. R. Stilwell, *J. Opt. Soc. Am.* (1941) *31*, 369–374.

4. H. F. Ives, *J. Opt. Soc. Am.* (1947) *37*, 810. Summary.

5. B. Rossi and D. B. Hall, "Variation of the rate of decay of mesotrons with momentum", *Phy. Rev.* (1941) *59*, 223.

6. R. P. Durbin, H. H. Loar, and W.W. Havens Jr., "The lifetimes of the π^- and π^+ mesons", *Phys. Rev.* (1952) *88*, 179; F. S. Crawford Jr., *Nature* (1957) *179*, 35–36. Summarizes experiments on mesons which verify time dilation. Also H. Dingle, *Nature* (1957) *179*, 865 and F. S. Crawford, *Nature* (1957) *179*, 1071.

7. R. Mössbauer, *Z. Physik* (1958) *151*, 124.

8. H. Hay, J. Schiffer, T. Cranshaw, and P. Egelstaff, *Phy. Rev. Lett.* (1960) *4*, 165. An extensive and recent review. Also F. Champion, *Nuovo Cimento* (1958) *7*, 122; and S. Raboy and C. C. Trail, *Nuovo Cimento* (1958) *10*, 797–803. The last two letters contain an argument and a counter-argument.

9. P. S. Farago and L. Janossy, *Nuovo Cimento* Series 10 (1957) *5*, 1411–1436.

10. P. L. Copeland and W. E. Bennet, *Elements of Modern Physics*, Oxford University Press, New York, 1961, pp. 60–72. Describes Perry and Chaffee's experiment on e/m_r for electrons by time of flight measurements.

11. S. Dushman, "Mass energy relation", *Gen. Elect. Rev.* (1944) *47*, 6–13. It gives graphs of values of and versus accelerating potential.

12. A. H. Compton, *Phys. Rev.* (1923) *21*, 207, 483; *Phil. Mag.* (1923) *46*, 897.

13. P. Debye, *Phys. Z.* (1923) *24*, 161.

14. A. H. Compton, *Phys. Rev.* (1923) *22*, 409; C. T. R. Wilson, *Proc. Roy. Soc. London.*

15. F. C. Champion, *Proc. Roy. Soc. London A* (1932) *136*, 630

16. Jorge Pullin and Zodolfo Gumbini, *Phys. Rev.* (1999)d59, 124021 H. Bravo-Alfaro, *Astronomical Journal*, 119, 580-592 (2000). J. R. Ellis et al, *Phys. Rev.* (1999) D59, 103503. J. R. Ellis, N. E. Mavromatos and D. V. Nanopoulos, *Solitons. Fractals* (1999a) 10, 345. S. M. Carrol et al, *Phys. Rev. Lett.* (2001087,141601). H. Mueller, *Phys. Rev.*(20050D71, 045004).

17. Alfrao et al., *AJ* (2000) *119*, 580–592.

18. Ellis et al., *Phys. Rev.* (1999) *D59*, 103503.

19. Ellis, Mavromatos and Nanopoulos, *Solitons Fractals*, (1999a) *10*, 345.

20. Calladay and Kostelecki, *Phys. Rev.* (1997) *D55*, 6760.

21. Carrol et al, *Phys. Rev. Lett.* (2001) *87*, 141601.

22. Muller, "Testing Lorentz invariance by the use of vacuum and matter filled cavity resonators", *Phys. Rev.* (2005) *D71*, 045004. A review article.

23. A. H. Compton, *J. Franklin Inst.* (1932) *136 A*, 630.

24. G. E. Uhlenbeck and S. A. Goudsmit, *Naturwiss* (1925) *13*, 953; *Nature* (1926) *117*, 264.

25. L. H. Thomas, *Nature* (1926) *7*, 514; *Phil. Mag.* (1927) *3*, 1.

26. W. H. Furry, "Lorentz transformation and Thomas precession", *Am. J. Phys.* (1955) *23*, 517–525, considers the kinematic effects of special theory of relativity by iteration of the infinitesimal Lorentz transformation.

27. W. Pauli, *Rapports du Septieme conseil de Physique Solvey*, Brussels 1933, Gauthier–Villars et cie, Paris, 1934.

28. F. Reines and C. L. Cowan, *Nature* (1956) *178*, 446.

29. T. D. Lee and C. N. Yang, *Phys. Rev.* (1956) *104*, 254.

30. C. S. Wu, E. Ambler, R. W. Hayward, D. D. Hoppes and R. P. Hudson, *Phys. Rev.* (1957) *105*, 1413.

31. *Proceedings of the Rehovoth Conference on Nuclear Structure*, H. J. Lipkion (ed.), North Holland Publishing Co., Amsterdam, 1958, pp. 376–403. A summary of the data as of 1957.

32. R. H. Dalitz, *Phil. Mag.* (1953) *44*, 1068; *Phys. Rev.* (1954) *94*, 1046.

33. N. N. Biswas et al., *Nouvo Cimento* (1956) *4*, 631.

34. T. D. Lee, and C. N. Yang, *Phys. Rev.* (1956) *104*, 254.

35. L. deBroglie, *Compt. Rend.* (1923) *177*, 507, 548, 630. Also refer R. Schlegel, *Am. J. Phys.* (1954) *22*, 77. Shows the equivalence of de Broglie equation and Newton's second law.

36. W. Heisenberg, *Z. fur. Phys.* (1927) *43*, 172.

37. The argument is due to Neils Bohr, cf. N. F. Mott and H. S. W. Massey, *Theory of Atomic Collisions*, Oxford University Press, 1949, Oxford, U.K.

38. R. L. Mössbauer, *Z. Physik* (1958) *151*, 124; *Naturwissenschaften* (1958) *45*, 538; *Z. Naturforsch.* (1959) 14*a*, 211. It was repeated by P. P. Craig et al., *Phys. Rev. Lett.* (1959) *3*, 221.

PROBLEMS

7.1 From the postulate of the equivalence of inertial frames, verify that $\mathbf{u}_a = \mathbf{u}_b$, where $\mathbf{u}_a$ is the velocity of X_2Y_2 as seen by XY and $\mathbf{u}_b$ is the velocity of XY as seen by X_2Y_2.

7.2 A point magnetic moment μ has a vector potential

$$\mathbf{A} = \frac{\mathbf{u} \times \mathbf{r}}{r^3}$$

and no scalar potential. Show that if the magnetic moment moves with a velocity v $(v < c)$, there is an electric dipole moment $\mathbf{p}$ associated with the magnetic moment, that is, $\mathbf{p} = \mathbf{v}/c \times \boldsymbol{\mu}$. Compute the interaction energy between the moving dipole and the fields **E** and **B**.

7.3 Show that the wavelength of electrons moving at a velocity very small compared with c and with a kinetic energy $\mathcal{E}_{kin}$ eV is given by

$$\lambda = \frac{12.268}{\sqrt{\mathcal{E}_{kin}}}\,\mathring{A}$$

Compute the wavelength of electrons with energies of 10, 50, and 100 eV.

7.4 Compute the wavelength of a thermal neutron taking the temperature equal to 27°C. The mass of the neutron is 1.008982 amu = $(1.67474 \pm 0.00010) \times 10^{-24}$ g.

7.5 When the wavelength λ of a neutron becomes comparable with the nuclear radius, the neutron can be diffracted about the nucleus to give what is known as shadow scattering. For a nucleus with radius equal to 5×10^{-13} cm, at what energy does this shadow scattering become important?

7.6 The position of a pointer in a measuring instrument is noted to within a precision of 0.1 mm. The mass of the pointer is 0.1 g. Find the uncertainty in the momentum of the pointer.

7.7 Modify the proof of Eq. (7.34) to remove the restriction $\langle z \rangle = \langle v \rangle = 0$.

7.8 For annihilation of an electron–positron pair at rest into a pair of photons, find the energy of photon and its frequency.
$(m_0c^2, m_0c^2/h)$

7.9 Show that a single photon in free space cannot give rise to pair production. However, with something nearby, e.g., a nucleus or a magnetic field, it can happen.

8

Introduction to General Relativity

General theory of relativity, a generalization of the special theory of relativity, is the geometric theory of gravitation. It was developed by Einstein[1] in 1916 with little or no experimental motivation, and motivated by philosophical questions. He was led to state that there is no absolute accelerated frame of reference, much the same way as there is no absolute velocity frame in the special theory of relativity. All accelerated motion is relative. Gravitation is a property of space–time, and the curvature of space–time is directly related to the four-momentum of the matter and radiation that is present.

Two years after the publication of *Special Theory of Relativity*, Einstein published a paper trying to modify Newton's theory of gravitation to fit relativity. Newton's expression for the gravitational attraction between two bodies depends upon the masses of the objects and the distance between the bodies assuming the action at a distance principle. The interaction being instantaneous travels faster than the velocity of light and thus is inconsistent with special relativity. The only inevitable conclusion that follows is that Newton's gravitational equation is not strictly true under circumstances when large velocities (and large masses) are involved.

While writing this paper a thought crossed Einstein's mind that he called the happiest moment of his life. In 1920, he commented as follows:

"*The gravitational field has only a relative existence. Because for an observer falling freely from the roof of a house-atleast in his immediate surroundings-there exists no gravitational field.*"

The predictions of general relativity differ markedly from those of classical physics, such as the geometry of space, the passage of time, propagation of light, and the motion of bodies in free fall. These differences include time dilation, the gravitational red-shift of light, and gravitational time delay. All these predictions have been amply verified experimentally. Although there are other relativistic theories of gravity, Einstein's theory is the simplest and consistent with experimental data. Nonetheless, there are problems as how to reconcile it with quantum physics to produce a self-consistent theory of quantum gravity.

Einstein's theory has astrophysical implications, like the existence of black holes—regions of space and time are distorted to an extent that nothing, not even light, can escape. The black holes are an end state of massive stars.

General relativity also predicts the existence of gravitational waves, which have since been measured indirectly and the projects like LIGO (Laser Interferometer Gravitational–Wave Observatory) and LISA (Laser Interferometer Space Antenna) aim at the direct measurement.

Furthermore, general relativity is the basis of the current cosmological models of the consistently expanding universe.

8.1 EQUIVALENCE OF INERTIAL AND GRAVITATIONAL MASS

The inherent properties of matter are inertia and gravitation. The gravitational mass of an object is a measure of that property of universal gravitation, on account of which every particle of matter attracts every other particle. Let M_g be the gravitational mass of the body, and one could measure it by measuring the force f exerted on it by another body, say the earth. Thus,

$$f = \frac{GM_E M_g}{r^2}$$

or

$$M_g = \frac{fr^2}{GM_E} \tag{8.1}$$

where G is the universal gravitational constant; M_E is the mass of earth; and r, the distance of the body from the centre of earth.

The inertial mass of an object relates to different property of matter, i.e., a force when impressed on the body will cause the body to accelerate. Thus, if M_i is the inertial mass, then

$$M_i = \frac{f}{a} \tag{8.2}$$

One is to check experimentally whether one can equate gravitational and inertial mass. The simplest experiment is to find whether all bodies, of different mass and composition, fall with the same acceleration. Consider a body of mass $M_i(1)$ and $M_g(1)$ falling near the surface of earth, then

$$M_i(1)\, a(1) = \frac{GM_E M_g(1)}{R_E^2} \tag{8.3}$$

Analogously, for a second body of mass, $M(2)$, one gets

$$M_i(2)\, a(2) = \frac{GM_E M_g(2)}{R_E^2} \tag{8.4}$$

Dividing Eq. (8.3) by Eq. (8.4), we get

$$\frac{M_i(1)\, a(1)}{M_i(2)\, a(2)} = \frac{M_g(1)}{M_g(2)}$$

or

$$\frac{M_i(1)}{M_i(2)} = \frac{M_g(1)\, a(2)}{M_g(2)\, a(1)} \tag{8.5}$$

Experimentally, it was demonstrated by Galileo that falling bodies always fall at the same rate so that $a(1) = a(2)$ and thus Eq. (8.5) becomes

$$\frac{M_i(1)}{M_i(2)} = \frac{M_g(1)}{M_g(2)} \tag{8.6}$$

Thus, experimentally there is a proportionality of gravitational and inertial mass. Experimentally, one is to examine whether there is any variation of the ratio M_i/M_g for different particles.

Newton showed that the difference between the gravitational and inertial mass is less than one part in 10^3 Bessel by calculating the time period of a pendulum as

$$T = 2\pi\sqrt{\frac{l}{g}\frac{M_i}{M_g}} \tag{8.7}$$

in lieu of the expression

$$T = 2\pi\sqrt{\frac{l}{g}} \tag{8.8}$$

which could push the accuracy to one part in 6×10^6. But Eötvös, through his elaborate experimentation over a period of 25 years, starting in 1890, could show that these masses are equal to within one part in 10^9. These experiments were based on measuring the direction of the freely hanging pendulum and thus the resultant of the forces f_i and f_g through a null experiment. Lastly, R. H. Dicke could further improve the accuracy of the equivalence of the gravitational and inertial mass to one part in 10^{10}.

8.1.1 Gravitational Mass of the Photon

A photon of energy $h\nu$ has inertial mass equal to $h\nu/c^2$. Pound and Rebka[2] showed that the inertial mass is equivalent to the gravitational mass of the photon, doing a Mössbauer[3] (nuclear γ-ray resonance) experiment on a γ-ray of 14.4 keV emitted by Co-57 radioactive source. By making the photon fall through a vertical height of 20 m, the shift in frequency was measured and compared with the value predicted by theory.

Let the photon energy $h\nu$ and frequency ν at a height H above the surface of earth be allowed to fall through the distance H, with the consequence that the energy increases by $m_g gH$ and now the energy is $h\nu'$. Hence

$$h\nu' = h\nu + \frac{h\nu}{c^2}gH \tag{8.9}$$

It was assumed that the mass was constant through the fall, which is reasonable if the change in frequency is not large. Thus,

$$\nu' = \nu\left(1 + \frac{gH}{c^2}\right)$$

or

$$\frac{\nu' - \nu}{\nu} = \frac{\Delta\nu}{\nu} = \frac{gH}{c^2} \tag{8.10}$$

Substituting $H = 20$ m we get

$$\frac{\Delta\nu}{\nu} = \frac{gH}{c^2}$$

$$\approx \frac{(10^3)\times(2\times10^3)}{(3\times10^{10})^2}$$

$$= 2\times10^{-15} \tag{8.11}$$

Pound and Rebka showed that

$$\frac{(\Delta\nu)_{exp}}{(\Delta\nu)_{theory}} = 1.5 \pm 0.10 \tag{8.12}$$

In a refined experiment subsequently, Pound and Snider[4] (1965) showed that

$$\frac{(\Delta\nu)_{exp}}{(\Delta\nu)_{theory}} = 0.9990 \pm 0.0076 \tag{8.13}$$

Thus the inertial and gravitational mass of the photon were shown to be equivalent.

8.2 THE PRINCIPLE OF EQUIVALENCE

The equivalence of the inertial and gravitational mass of a body has the implication that the gravitation in a sense is equivalent to acceleration. It follows from the equivalence principle formulated by Einstein in 1911. It is stated as follows:

"*The effects produced by gravitational field are identical to those due to accelerated motion and are completely indistinguishable from each other.*"

Accordingly, an observer in a freely falling elevator will find the laws of physics to be the same as in inertial frames. Furthermore, an observer in a satellite going around the earth will experience weightlessness, since the gravitational force gets cancelled by the orbital acceleration $a = \omega^2 r = GM/r^2$, where r is the distance from the centre of earth to the satellite.

Einstein was led to state that as there is no absolute inertial frame in special relativity, there is no absolute accelerated frame of reference. All accelerated motion is relative.

The principle of equivalence, like the principle of relativity, was in no way a new idea when Einstein in 1915 made it as one of the postulates of general relativity. Aristotle did not accept it but contrarily believed that heavy objects fall faster than the light ones. But Galileo demonstrated it publicly by dropping unequal masses from the leaning tower of Pisa and he studied it by rolling down an inclined plane cylinders of different masses and composition. The accuracy of measurement was refined successively by Newton, Bessel, Eötvös, and Dicke, as described earlier in Section 8.1.

8.3 EINSTEIN ELEVATOR

This was a thought experiment by Einstein to illustrate the implications of the principle of equivalence. Consider an inertial frame S, deep in interstellar space, far from any source of gravitational field. To the observer located in S, the masses do not seem to have accelerations. Consider another frame of reference S' that is uniformly accelerated with respect to S. However, to the observer in S', all the masses appear to have equal and parallel accelerations. If the observer in S' is not aware of the acceleration of S' with respect to S, he will infer that the acceleration of S' is due to a uniform gravitational field. He could verify his inference by studying the time period of a simple pendulum. The viewpoint of observer in S' is equally valid as that held by the observer in S.

To put it mathematically, we assume that at $t = 0$ the origins of S and S' coincide and the system S' has acceleration a in the negative direction of z-axis. Thus,

$$\mathbf{r} = \mathbf{r}' - \frac{1}{2} a t^2 \mathbf{k} \tag{8.14}$$

or
$$\ddot{\mathbf{r}} = \ddot{\mathbf{r}}' - a\mathbf{k} \tag{8.15}$$

or
$$m_i\ddot{\mathbf{r}} = m_i\ddot{\mathbf{r}}' + m_i a\mathbf{k} \tag{8.16}$$

If the impressed force in S is f and in S' it is f', then

$$f' = f + m_i ak \tag{8.17}$$

Thus we see that in S' in addition to the impressed force f, there is present an inertial force $m_i ak$. If the acceleration $a = g$, the acceleration due to gravity, then the inertial force cancels the gravitational force. The observer in S' will conclude that no force of gravity is operating at all.

Now assume that S' is removed away from all other objects and is moving upwards along the positive direction of z-axis, with acceleration a. Then

$$\mathbf{r} = \mathbf{r}' + \frac{1}{2}at^2\mathbf{k} \tag{8.18}$$

and
$$m_i\ddot{\mathbf{r}} = m_i\ddot{\mathbf{r}}' - m_i a\mathbf{k} \tag{8.19}$$

$$\Rightarrow f' = f + m_i ak \tag{8.20}$$

The observer in S' will conclude that every mass is under a gravitational force equal to $m_i ak$ which gives an acceleration equal to $m_i a/m_g$. Since the inertial and gravitational masses are equal, so $a = g$. Furthermore, the observer in S' will infer that the absence or presence of gravitational field is entirely dependent on the choice of the frame of reference used to describe the physical phenomenon. Alternately stated, the gravitation is a property of the space–time continuum.

Thus it is on the basis of the principle of equivalence that we realize the unity of inertia and gravitation: It all depends how one looks at the same masses that may appear to be either under the influence of inertia alone (as in system S) or under the combined action of inertia and gravitation (as in system S'). Thus there is no difference between the inertial system with gravitational field or a non-inertial system.

A freely falling observer in a uniform gravitational field experiences weightlessness. One can state the principle of equivalence in an alternate manner: "It is always possible in the case of a homogeneous gravitational field to transform to a new set of space-time coordinates, where there is no gravity."

In the case of inhomogeneous gravitational field, the acceleration due to gravity will be different at different points. Under such circumstances, the principle of equivalence has to be modified. We can maintain that for a sufficiently small region of space, the effect of gravity can be annulled by choosing a frame of reference that is falling freely with an acceleration that is equal to that due to gravity in that region; for example, the effect of gravity is eliminated in a freely falling chamber at the surface of earth. In the light of discussion so far, one could state the principle of equivalence as follows:

"*At any arbitrary point in space-time, it is always possible to transform to a coordinate system such that the effects of gravity are eliminated in the small differential region around that point.*"

8.4 THE PRINCIPLE OF COVARIANCE

If the principle of covariance is to be built in as the integral part of the theory of gravitation, then the concept of inertial frames of reference has to be abandoned. In general relativity, we incorporate all kinds of motion and there is no reason to choose an inertial frame of reference over a non-inertial frame. In order to make use of the general idea of relativity, encompassing all kinds of motion, it is imperative that the laws of physics are expressed in a form which is independent of the choice of the coordinate system, or in other words the laws of nature remain invariant with respect to any space–time coordinate system. This statement is called the *principle of covariance.* This is achieved through the use of tensor calculus, since the expression of a law by a tensor equation preserves its form in all systems of coordinates.

In special theory of relativity, we described the basic laws of physics in a four-dimensional space–time geometry described by the line element

$$ds^2 = c^2dt^2 - dx^2 - dy^2 - dz^2 \tag{8.21}$$

Equation (8.21) remains invariant for the Lorentz transformations from one Cartesian coordinate system (ct, x, y, z) to another coordinate system (ct', x', y', z'), which is moving in uniform relative motion with respect to the other. However, under transformation to either an accelerated frame of reference or curvilinear coordinate system, Eq. (8.21) will not remain invariant. Thus to be in conformity with the principle of covariance, the required expression for the line element for specifying the space–time geometry has to be a covariant expression. Only then, the space–time continuum can be the basis for the development of the general theory of relativity.

The line element, Eq. (8.21), can be put in a tensor form

$$\begin{aligned} ds^2 &= g_{\mu\nu}\, dx_\mu\, dx_\nu \\ &= (dx^0)^2 - (dx^1)^2 - (dx^2)^2 - (dx^3)^2 \end{aligned} \tag{8.22}$$

where $x^0 = ct, x^1 = x, x^2 = y, x^3 = z$, and $g_{\mu\nu}$ is the metric tensor with constant components

$$g_{\mu\nu} = \begin{pmatrix} 1 & 0 & 0 & 0 \\ 0 & -1 & 0 & 0 \\ 0 & 0 & -1 & 0 \\ 0 & 0 & 0 & -1 \end{pmatrix} \tag{8.23}$$

Further we consider a general coordinate transformation to a new set of axes which are not necessarily Cartesian. Let the new axes be chosen as $x'^\mu = (x'^0, x'^1, x'^2, x'^3)$

$$x'^\mu = x'^\mu\,(x'^0, x'^1, x'^2, x'^3) \tag{8.24}$$

The differentials dx^μ in terms of dx'^μ are

$$dx^\mu = \frac{\partial x^\mu}{\partial x'^\alpha}\, dx'^\alpha \tag{8.25}$$

The line element in terms of the new coordinates dx'^{μ} becomes

$$ds^2 = g_{\mu\nu}(x)\, dx'^{\mu}\, dx^{\nu}$$

$$= g_{\mu\nu}(x)\left(\frac{\partial x^{\mu}}{\partial x'^{\alpha}}\, dx'^{\alpha}\right)\left(\frac{\partial x^{\nu}}{\partial x'^{\beta}}\, dx'^{\beta}\right)$$

$$= g_{\mu\nu}(x)\,\frac{\partial x^{\mu}}{\partial x'^{\alpha}}\frac{\partial x^{\nu}}{\partial x'^{\beta}}\, dx'^{\alpha} dx'^{\beta}$$

$$= g'_{\alpha\beta}(x')dx'^{\alpha}\, dx'^{\beta} \tag{8.26}$$

where
$$g'_{\alpha\beta}(x') = g_{\mu\nu}(x)\frac{\partial x^{\mu}}{\partial x'^{\alpha}}\frac{\partial x^{\nu}}{\partial x'^{\beta}} \tag{8.27}$$

Equation (8.27) is the law of transformation for a covariant tensor of rank two. Thus the covariant expression for interval with respect to the general coordinate transformation is

$$ds^2 = g_{\mu\nu}(x)\, dx^{\mu}dx^{\nu} \tag{8.28}$$

As is obvious, the components of the metric tensor $g_{\mu\nu}$ are no longer constants but are functions of the coordinates. The line element of the special relativity is only a particular case of Eq. (8.28).

8.5 GRAVITATIONAL EFFECT ON CLOCKS

In the vicinity of massive objects, clocks run slowly than they do in free space. We will derive this change of rate of a clock in a gravitational field, making use of the mass–energy relationship and the principle of conservation of energy. A photon of mass m, rising through a gravitational potential difference $gH = \frac{GM_s}{R_s}$, where g is the gravitational acceleration and H is the height, will lose an amount of energy mgH. Unless the photons lose energy mgH, the energy is not conserved.

The energy of the photon is $\mathcal{E} = h\nu = mc^2$ and the change of energy is $\Delta\mathcal{E} = mgH = h\Delta\nu$.

Hence,
$$\frac{\Delta\mathcal{E}}{\mathcal{E}} = \frac{\Delta h\nu}{h\nu} = \frac{mgH}{mc^2}$$

$$\Rightarrow \frac{\Delta\nu}{\nu} = \frac{gH}{c^2} \tag{8.29}$$

The same result is derivable using the principle of equivalence, and Doppler effect. We consider an accelerating spaceship with one clock at its front and the other at its rear and try to find the difference in frequency. During the time a photon from the front of the spaceship reaches the rear, the ship's speed has increased and as a consequence, the frequency of the photon has increased slightly from the frequency when the photon was transmitted from the front.

Assuming that the height of the ship is H, the time taken by the photon to reach the rear is $t = H/c$. During the time t, the increase in the speed of the ship is $v = at = aH/c$. For small β the Doppler equation is $\nu' = \nu(1 + \beta)$ or $\Delta\nu/\nu = \beta$. Thus we find that

$$\frac{\Delta\nu}{\nu} = \frac{aH}{c^2} \tag{8.30}$$

By the principle of equivalence, the same effect will occur in a gravitational field that produces an acceleration g, so

$$\frac{\Delta\nu}{\nu} = \frac{gH}{c^2} \tag{8.31}$$

which is the same as the previous result.

The general theory predicts that the solar spectral lines will undergo a gravitational 'red shift' or decrease in frequency of about $\Delta\nu/\nu = 2 \times 10^{-6}$, due to the large negative gravitational potential of the sun's enormous mass. This was confirmed experimentally by J. Brault in 1962.

Due owing to their large negative gravitational potential, dwarf stars have large values of the gravitational red-shift. For Sirius B, the calculated fractional shift is

$$\frac{\Delta\nu}{\nu} \cong 5.9 \times 10^{-5} \tag{8.32}$$

which compares not too unfavourably with the observed value of -6.6×10^{-5}. The disagreement arises from the uncertainties in M_s and R_s, since $gH = \frac{GM_s}{R_s}$.

The effect was observed terrestrially by Pound and Rebka (1960) using the Mössbauer effect, as discussed in detail in Section 8.2.1.

8.6 CLASSICAL TESTS OF GENERAL RELATIVITY

After the publication of the general theory of relativity, there was a vigorous drive to test its predictions. Einstein, on the basis of his theory, could explain a small discrepancy in the motion of the perihelion of planet Mercury. However, the most dramatic verification of the bending of starlight by Sun, by British astronomers in 1919, made Einstein a world celebrity overnight. Since 1916, there have been many verifications of the diverse predictions of general relativity, but we will concentrate on the 'classical' tests of the theory considered by Einstein:

1. The gravitational red-shift
2. Solar bending of starlight
3. Precession of the perihelion of Mercury

8.6.1 The Gravitational Red-Shift

The red shift considered in Section 8.5 has been observed both astronomically as well as terrestrially. It is a test of that part of general relativity, which is based on the principle of equivalence.

8.6.2 Solar Bending of Starlight

Soldner in 1801 predicted the solar deflection of starlight. Light was considered to have particle nature before Young's experiment in 1802 showing the interference of light and thereby demonstrating the wave characteristics of light. Again light was ascribed the particle character, but in somewhat different sense, after Einstein's theory of photoelectric effect in 1905.

Considering that light consists of photons of mass m, travelling at velocity c, it will be deflected by $2GM/bc^2 = 0.87$ second of arc, nonrelativistically, where G is the universal gravitational constant, M is the mass of the Sun, and b is the distance of closest approach of the photon to the Sun. It may be noted that the mass of the photon does not enter into the expression for deflection.

In 1911, Einstein by a different argument arrived at the same result of 0.87 second of arc. As discussed in an earlier article, a clock will go slower when placed in a low gravitational potential which exists near the Sun. If the clock is a light-clock and it runs slow, then assuming that transverse distances are not affected by the gravitational field, the light velocity has to be reduced according to the rate of the clock. As light is bent in passing into a medium, such as glass in which it moves slower than in vacuum, in much the same way, light gets bent near the Sun, as it moves more slowly.

In 1915, Einstein included the affects of the curvature of space–time caused by the presence of a massive body like the Sun and this resulted in doubling the effect to 1.75 seconds of arc.

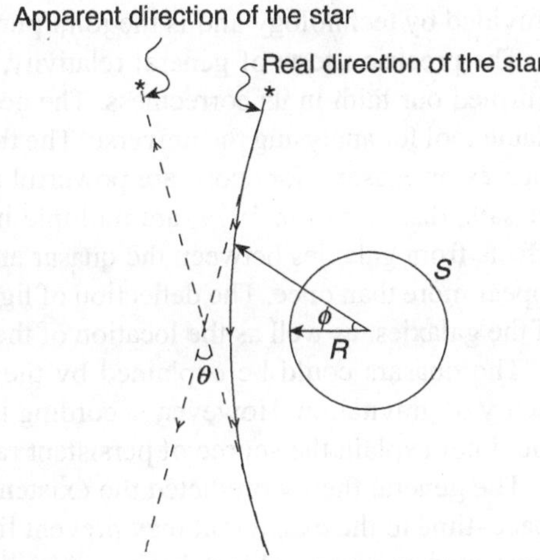

Figure 8.1 *Bending of starlight near the Sun.*

The bending of starlight is difficult to observe, since the starlight passing close to the sun is visible only during a total eclipse of the Sun. In 1919, Arthur Eddington observed that such a deflection of starlight as it grazed the Sun does take place, Figure 8.1. The measurement in agreement with the prediction of general relativity helped in making Einstein a world celebrity. The observations were made simultaneously in the city of Sobral, Ceara, Brazil and in the west coast of Africa. The result was considered spectacular news and made the front page of most newspapers.

When asked by his assistant what his reaction would have been if general relativity had not been confirmed by Eddington and Dyson[5] in 1919, Einstein famously made the quip, 'Then I would feel sorry for the dear Lord. The theory is correct anyway.'

The early accuracy, however, was poor. Dyson et al. quoted optimistically low uncertainty in their measurements, which is argued by some to have been plagued by systematic error and probably confirmation bias, although modern reanalysis was correct.

8.6.3 Precession of the Perihelion of Mercury

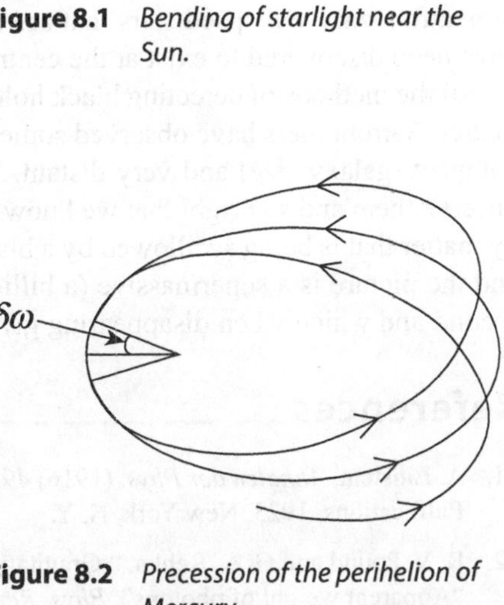

Figure 8.2 *Precession of the perihelion of Mercury.*

The elliptical orbit of the planet Mercury is slowly rotating with respect to the fixed stars and the total precession is 5,600 seconds of arc per century, Figure 8.2. A careful calculation based on Newton's theory of gravitation, including the effects of perturbation due to other planets in the solar system, predicted a precession of 5,557 seconds of arc per century. The precision of astronomical measurements is so great that a difference of 43 s is hundred times the observation error. Einstein's calculation on the basis of the general relativity gave the additional 43 seconds, thus providing the quantitative proof of the correctness of general relativity.

8.7 THE RENAISSANCE OF GENERAL RELATIVITY

By the 1950s, the interest in general relativity had cooled off, since it lacked observational evidence of its predictions, in addition to having several problems and ambiguities. However, in 1959, there was a resurgence in general relativity research due to the discovery of a method for performing more accurate tests of the theory. It was the radar echo from Venus that was used to carry out high precision tests of general relativity, since small angular deflections of light and sound by the Sun could be measured with great precision. This was a tremendous improvement over the experiments of British astronomers, carried out in 1919. This renewed interest continues to be unabated due to better instrumentation provided by technology and is the joint pursuit of many ongoing international collaborations.

The precision tests of general relativity, carried out in 1960s and 1970s, supported the theory and affirmed our faith in its correctness. The general theory is not only a theory of space and time but is a viable tool for analysing the universe. The theory was in the study of double quasars. Quasi-stellar radio sources or quasars, for short, are powerful sources of radio waves from very distant galaxies. Double quasars, discovered in 1979, are multiple images of the same quasar, caused by gravitational lensing effects from galaxies between the quasar and the observer, since it causes the light from the quasar to appear more than once. The deflection of light from the quasar, the characteristics and mass distribution of the galaxies, as well as the location of the quasar are analysed on the basis of general relativity.

The quasars could be explained by the existence of supermassive stars according to Newtonian theory of gravitation. However, according to general theory, the supermassive stars were unstable and could not explain the source of persistent radio waves.

The general theory predicted the existence of black holes. Black holes are so dense that they bend space–time to the extent that they prevent light from escaping. As light, which is the fastest in the universe, cannot escape a black hole, nothing else is able to escape either. General theory limits the mass of neutron stars, collapsed stars that are black hole candidates. Black holes are no longer elusive and have been discovered to exist at the centre of some galaxies.

All the methods of detecting black holes depend on the manner in which they affect the surrounding matter. Astronomers have observed some very distant x-ray sources, which are known to be relatively compact (galaxy size) and very distant. These are very bright sources and that is the reason we can observe them and so bright that we know of only one source that can fuel them, the radiation given off by matter that is being swallowed by a black hole. These bright objects are called active galactic nuclei and the picture is a supermassive (a billion solar masses or so) black hole, devouring many stars per second and which when disappearing give off the energy that announces their demise.

References

1. A. Einstein, *Annalen der Phys.* (1916) *49*, 769. For English translation, see *The Principle of Relativity*, Dover Publications, 1923, New York, N. Y.

2. R. V. Pound and G. A. Rebka, " Gravitational red shift in nuclear resonance", *Phys. Rev. Lett.* (1959) *3*, 439; "Apparent weight of photons", *Phys. Rev. Lett.* (1960) *4*, 337.

3. R. L. Mössbauer, *Z. Physik* (1958) *151*, 124; *Naturwissenschaften* (1958) *45*, 538; *Naturforsch* (1959) *14a*, 211.

4. R. V. Pound and J. L. Snider, "Effect of gravity on gamma radiation" *Phys. Rev.* (1965) *140B*, 788.

5. F. W. Dyson, A. S. Eddington, and C. Davidson, "A determination of the deflection of light by sun's gravitational field from observations made at the solar eclipse of May 29, 1919", *Phil. Trans. Roy. Soc. A 220*(571–581), 291–333, (1920).

Problems

8.1 The best statement of equivalence principle is:

(a) The laws of physics are the same in an accelerated reference frame and in a uniform, static gravitational field.
(b) Tidal forces cannot be detected.
(c) Inertial and accelerated reference frames cannot be differentiated.

8.2 A clock at the surface of a star runs slow by 30%. Calculate the radius of the star in kilometres and light years; surface gravity = 1,000 cm/s^2. $(2.7 \times 10^{12}$ km; $9/10\pi$ light years$)$

8.3 Calculate the change in wavelength of light ($\lambda = 6{,}000$ Å) falling 60 ft in the gravitational field of earth. $(1.2 \times 10^{-13}$ Å$)$

8.4 Compare the solar gravitational red-shift with the Doppler shift at a point at the edge of the sun, on its equator, which is moving directly towards the earth. The sun is about a million miles in diameter $(r = 7 \times 10^8$ m$)$ and rotates every 25 days. Its mass M is 2×10^{30} kg. The Gravitational constant is $6\ 7 \times 10^{-11}$ N m^2/kg^2. Use a gravitational potential of $-GM/r$ instead of gh, since the field is not constant. Neglect the motion and the gravitational potential of the earth. (Gravitational $\dfrac{\Delta \nu}{\nu} = -2.1 \times 10^{-6}$; Doppler $\dfrac{\Delta \nu}{\nu} = 6.8 \times 10^{-6}$)

Bibliography

This is a list of useful works. References to the original papers on topics of specialized interest are given in the References section at the end of each chapter.

Introductory Textbooks on Special Theory of Relativity

Das, A., *The Special Theory of Relativity: A Mathematical Approach,* Springer-Verlag Berlin Germany, 1996.

Naber, G. L., *The Geometry of Spacetime: An Introduction to the Mathematics of Special Theory of Relativity*, Springer-Verlag, Berlin Germany, 1992.

Rindler, W., *Introduction to Special Relativity*, 2nd edn., Oxford University Press, 1991.

Taylor, E. F. and J. A. Wheeler, *Spacetime Physics: Introduction to Special Relativity,* 2nd edn., W. H. Freeman & Company, New York, N. Y. 1992.

Woodhouse, N. M. J., *Special Relativity* (Springer Undergraduate Mathematics Series), Springer, Berlin Germany, 2007.

Index

aberration, 17, 18, 36, 121–122
absolute ether frame, 8
absolute space, 5–6
  time, 5–6
ammonia masers, 13–15
angle of scattering, 113
anti-proton, production of, 107, 108
anti-symmetric four-tensor, 65–68

base vectors, 49
body alpha, 3

causality, 33
centre of momentum transformation, 106–107
charge density, 63
classical tests of general relativity, 212–214
clock hypothesis, 32
clocks
  apparent slowing of, 31–32
  gravitational effect on, 211–212
C-M system, 106–107
collision cross-section, 108
  kinematic relations, 110
Compton effect, 100
conservation laws of vacuum
    electrodynamics, 163–165
conservation of momentum, 81–88
constancy of velocity of light, 18
covariant formulation of field
  conservation laws, 163–165
  Maxwell's equations, 144–147
  of Lorentz condition, 64–67

D' Alembertian operator, 26, 132
de-Broglie waves, 194
divergence of a vector, 52
Doppler effect, 121–126
  longitudinal, 123
  transverse, 123

Einstein elevator, 208–209
Einstein, special relativityand Newton's laws, 18
electromagnetic energy-momentum
    tensor, 160–165
electromagnetic field tensor, 137–140
  electromagnetic field invariants, 140–141
electromagnetic waves, 7, 19
energy of a free particle, 76–80
  photons, 103–104
energy-mass equivalence, 94–99
equation of continuity, 64
equation of motion of a charged
    particle, 151–160
equivalence principle, 208
ether, 7
  ammonia masers, 13–15
  drift, 13, 15
  frame, 9, 17
Euclidean manifold, 55
event, defined, 20

field of a charge in uniform motion, 165–168
field strength of an accelerated charge, 171
Fizeau's experiment, 35
four tensor, 65–67
four vector current, 62
four vectors, 60
four-dimensional acceleration, 88–89
  momentum, 77
  velocity, 88
frequency transformation
    of electromagnetic waves, 120

Fresnel's dragging coefficient, 126
fundamental equation
  relativistic dynamics and, 90

Galilean relativity, 2
  transformation, 2
gauge transformation, 132
gravitational mass, 206–208
  of photon, 207–208
gravitational red shift, 207–208

Hamilton's canonical
  equations, 150
Hamiltonian for a charged particle,
  in electromagnetic field, 148–149
Hamiltonian of a free particle, 79–80
historical perspective of relativity, 7

inertial forces, 5
inertial frame, 5
inertial mass, 82, 206–207
interval, 24, 40–41
  light-like, 41
  space-like, 41
  time-like, 41
invariance
  under Lorentz transformation 131–134
  of velocity of light, 20

Kennedy–Thorndyke experiment, 10
  modern re-enactments of, 11
kinematic relations for collisions, 110

Lagrangian for a free particle in
  electromagnetic field, 76
Lande's splitting factor, 184
Larmor precession frequency, 184
law of composition of velocities, 39, 40
  inertia of energy, 94
laws of inertia of energy, 94
length contraction, 30–31
  by a single observer, 30
Lienard–Wiechert potentials, 170–171
light cone, 40–41
linear orthogonal transformation, 49–53
longitudinal Doppler effect, 123

Lorentz condition, 64
  covariant form, 64
  equation, 64
  frames, 24
  group, 69
  invariance of continuity
  invariant, 61, 81
Lorentz condition, 64, 133
Lorentz contraction factor, 23
Lorentz invariance, 183
Lorentz transformation, 23–24, 70
  axiomatic derivation, 20–24
  of field components, 134, 141
  geometrical interpretation, 37–40
  group property, 68–73
  inverse, 23–24
  kinematic consequences, 28–31
  as rotation in four-space, 55–60
Lorentz–Fitzgerald contraction, 28–31

maximum signal velocity, 33
Maxwell's field equations
  by variational technique, 157–160
  covariant form, 144–147
mechanics, relativistic, 75–76
  momentum in, 76
mercury, 213–214
Michelson–Morley experiment, 8–10
Minkowski force, 90–93
Minkowski world, 54
momentum in relativistic mechanics,
  76, 81–84, 114
  energy four-vector, 62
moving clock, apparent slowing of, 31–32
moving faster than light, 193–194

Newton's laws, 4–5
Newtonian mechanics, 4, 63, 76, 81
non-conservation of parity, 191–193
null four-vector, 61

orthochronous Lorentz group, 73
orthogonality condition, 50

parity non-conservation, 191
phase velocity, 195

π^0 mesons, decay of, 103
$\pi^\pm$ mesons, decay of, 97
pilot waves, 194
point charge, in uniform motion, 165
potential generalized, 149
Poynting vector, 164
precession of perihelion of mercury, 213–214
pre-relativity physics, difficulties in, 1–4
principle of covariance, 210–211
principle of equivalence, 208
principle of relativity, 18
principle of the constancy
  of Galilean relativity, 4
  of velocity of light, 18
proper length, 28
proper mass, 86
proper time, 31
pseudo-Euclidean space, 55

Q-value, 95

reaction threshold, 107
reciprocal transformation, 108
red shift of spectral lines, 123, 212
relative velocity, 44
relativistic Doppler effect, 121–126
  classification of particles, 188–190
  dynamics, equation of, 90–92
  Hamiltonian of a free particle, 79
  Langrangian for a free particle, 75–76
  mechanics, 75
relativity of simultaneity, 32
renaissance of general relativity, 214
rest mass energy, 94
retarded potentials, 168–170

signal velocity, 4
signature of space, 55
slowing of moving clocks, 31–32
solar bending of starlight, 212–213
solid angle transformation, 124
special theory of relativity
  postulates, 17–18
special theory of relativity, experimental verification, 179–188
  motivation for testing, 182–183
  testing the consequences, 180
  testing the postulates, 179–180
spectral lines, red-shift, 123, 212
spin magnetic moment, 184
spin orbit interaction, 183–188

θ-τ puzzle, 193
tensor algebra, 65–68
tensor components, 65
Thomas precession, 183–188
Thomason velocity selector, 156
time dilation, 31–32
  experimental verification, 31
transformation equation
  acceleration, 42–45
  electromagnetic field, 134–137, 141–144
  force, 92–93
  of a four vector, 63–64
  four vector potential, 62, 133
  Lorentz contraction factor, 23–24, 43
  mass, 86–88
  velocity, 42–45
transformation of force, 92–93
transverse Doppler effect, 123–126
Trouton and Noble experiment, 12–13

uncertainty principle, 197
unimodular transformation, 73

variant scalar, 51
variation of mass with velocity, 81–89
vector potential, 119, 120, 131
velocity addition theorem, 33–35
velocity of interaction, 4
velocity, addition of, 33

wave equation under Galilean
    transformation, 6–8
wave propagation, elements in, 117–120
wave-vector transformation, 120
world line, 38–39
world point, 38–39

zero rest-mass particle, 190–194
zero-momentum reference system, 110